CRC
Handbook of
VIRUSES

Mass-Molecular
Weight Values
and
Related Properties

Horace M. Mazzone

CRC Press

Boca Raton Boston London New York Washington, D.C.

Acquiring Editor: Harvey Kane
Project Editor: Joanne Blake
Marketing Manager: Becky McEldowney
Cover design: Dawn Boyd
PrePress: Kevin Luong

Library of Congress Cataloging-in-Publication Data

Mazzone, Horace M.
    CRC handbook of viruses : mass-molecular weight values and related
properties / Horace M. Mazzone.
       p.   cm.
    Includes bibliographical references and index.
    ISBN 0-8493-2625-7 (alk. paper)
    1. Molecular virology--Handbooks, manuals, etc.   2. Viruses-
-Handbooks, manuals, etc.   I. Title.
QR389.M39   1998
579.2--dc21
                                              98-2701
                                              CIP

No claim to original U.S. Government works
International Standard Book Number 0-8493-2625-7
Library of Congress Card Number 98-2701
Printed in the United States of America 1 2 3 4 5 6 7 8 9 0
Printed on acid-free paper

Preface

My professional interest has centered on viruses, and the molecular weight and related properties of these entities have been a major concern. The objective of this handbook is to acquaint the investigator with a number of methods that can be used to obtain the mass-molecular weight value of viruses. No preference of methods is intended. Rather, if access to suitable equipment is possible, "old or new," molecular weight studies can be undertaken. In this connection I sincerely hope this handbook will serve as a source of information.

A number of excellent publications, books and original articles, were consulted in writing the various chapters. These works should be noted by the reader. The chapters were developed based on the following rationale: Chapter one serves to discuss viruses as infectious agents and traces their role in establishing the relatively new discipline of molecular biology. In Chapter two, the essential components of viruses, protein and nucleic acid, are discussed. I considered their discovery, nature, and structural organization, and how they join as nucleoproteins to form viruses. In Chapter three, some basic aspects of virus purification are noted. At some point, the investigator will need to prepare purified and concentrated fractions of the virus of interest. Usually, this will be done with a preparative centrifuge, and various purification methods are given.

Chapter four presents basic aspects of crystallography. Among the first molecular weight values reported for viruses were those estimated through crystallographic procedures. The X-ray analysis of the viral protein component leads to knowledge of how subunits are arranged and insight on how the intact virus is composed. Crystallography is the logical adjunct of molecular weight studies.

Chapters five and six constitute the core chapters of the book. These treatises explain and discuss and give examples of the various sedimentation experiments and scattering studies used to obtain the molecular weight. For sedimentation velocity experiments, the sedimentation and diffusion coefficients, integral to the basic Svedberg equation, are considered in Chapter five. The notable time-saving feature of obtaining the diffusion coefficient of viruses through dynamic light scattering is endorsed. With this facility the diffusion coefficient, in this investigator's opinion, should be restored as a basic property of viruses, and noted as such in the classification of these infectious agents. A number of sedimentation equilibrium procedures are also presented in Chapter five. The approach to equilibrium method, which had fallen into disuse, has a good chance for revival with the implementation of computers and instrument upgrading, as witnessed in the design of the new Beckman Optima series of analytical ultracentrifuges.

Scattering studies, discussed in Chapter six, include small angle X-ray, small angle neutron, classical light scattering, and electron microscopy. A rebirth of instrumentation, notably in classical light scattering techniques, and the availability, particularly, of the cold neutron facilities currently being brought into operation by government laboratories, should stimulate investigators to reconsider scattering procedures to obtain mass values and molecular weights of viruses.

Chapter seven serves, it is hoped, to stimulate interest in the sizing and solvation of viruses in solution, their "natural environment." These important considerations augment one's understanding of the molecular weight value, and the modeling experiments discussed are crucial in this connection.

Chapter eight, Available Resources, is an important resource guide for investigators lacking facilities or instrumentation required to pursue molecular weight studies. In my own experience, the ability to use the high voltage transmission electron microscope in order to obtain mass values of viral inclusion bodies was a major satisfying endeavor. This research resource was made available to me through the U.S. government's National Center for Research Resources, an important helping arm of the National Institutes of Health.

Acknowledgments

This book became possible through the support of a number of individuals and institutions. My introduction to virus research occurred when the U.S. Army placed me in the ranks of the Korean Hemorrhagic Virus Research Unit. Later, as a student at the University of Wisconsin, I was truly fortunate to be guided by dedicated teachers: J.N. (Jerry) Williams, Jr., Paul Kaesberg, my thesis advisor, who introduced me to electron microscopy, Robert Bock, whose knowledge of density gradient systems allowed me to get a hold on my virus research problem, and Robert L. Baldwin, an expert in sedimentation analyses. At Wisconsin, I became familiar with the Model E Analytical Ultracentrifuge in obtaining the molecular weight and other properties of a plant virus.

As a postdoctoral student, various investigators, by way of their expertise in virology, helped me to understand the various fascinating aspects of viruses: at the Cold Spring Harbor Laboratory, Dr. A.D. Hershey; at the Harvard Medical School, Dr. Gordon Julian; at the Massachusetts Institute of Technology, Professor Cecil Hall and Dr. Henry Slayter; along with Professor Karl Schmid (Boston University Medical School).

During twenty years as a microbiologist with the U.S. Department of Agriculture, I had the opportunity to engage in collaborative virus research with a number of gifted investigators: Drs. Norman Anderson and Julian (Bud) Breillatt of the Oak Ridge National Laboratory; Dr. Gunter Bahr and Walter Engler at the Armed Forces Institute of Pathology; Dr. William Bancroft at the Walter Reed Army Institute of Research; Dr. Robert Fisher, Al Szirmae, and Bob Conroy at the U.S. Steel Research Center; the late Professor Keith Porter and George Wray at the University of Colorado; Drs. Ed Uzgiris and Ralph DeBlois at the General Electric Research Center; the late Professor J. Calvin Giddings and Dr. Karen Caldwell at the University of Utah.

In writing the various chapters of this book, a debt of gratitude is owed to a number of individuals: for sedimentation analyses, Ken Johnson of Beckman Instruments, as well as the Beckman Instruments library staff; for light scattering studies, Dr. David Nicoli, Particle Sizing Systems, and Cliff Wyatt of Wyatt Technologies; for crystallographic information, Bob Cudney of Hampton Research for calling my attention to the NIST/NASA/CARB BMCD Data Base on Virus Crystal Studies, and Gale Rhodes, University of Southern Maine; for preparation of drafts, Kathy Johnson and Lisa Bartlett; for assistance in computer processing, Darrin Snihur.

I am especially grateful to CRC Press for supporting this endeavor.

Dedication

To Arlene and Anne, and to the memory of my parents.

Horace M. Mazzone

The author, Horace M. Mazzone, Ph.D., first became interested in viruses as a member of a research team studying hemorrhagic fever virus during the Korean War. During his doctoral (University of Wisconsin) and postdoctoral years (Cold Spring Harbor Laboratories, Harvard Medical School, and the Massachusetts Institute of Technology), Dr. Mazzone studied various classes of viruses with a deep concern for elucidating the properties of these infectious agents, with special reference to the mass-molecular weight value.

In addition to obtaining the molecular weights of many viruses through sedimentation analyses employing sedimentation and diffusion coefficients, and sedimentation equilibrium studies, Dr. Mazzone has also used sedimentation field flow fractionation to obtain such values. Molecular weight studies on larger viruses were made possible through the use of the transmission electron microscope. Dr. Mazzone and colleagues published, in a series of articles, the first report of the mass of insect viral inclusion bodies, obtained with the high voltage electron microscope.

Retired from the U.S. Department of Agriculture, Dr. Mazzone has maintained his interest in viruses in addition to his teaching duties as an adjunct professor at Norwalk Community-Technical College, Norwalk, Connecticut. A future project is the preparation of a database of viruses in terms of their physical properties.

Contents

1

Introduction

Demonstrating properties of life only when activated in living cells, viruses as infectious agents have posed an intriguing study since their discovery about 100 years ago. In the host cell viruses multiply, adapt, mutate, and undoubtedly evolve.

In retrospect, a number of observations on the effects of viral diseases were noted by Sir F. C. Bawden, in his classic work, *Plant Viruses and Virus Diseases*.[1] In the 16th century, tulips infected with the tulip mosaic virus were observed to change their color. This condition, referred to as "color breaking," featured prominently in many paintings by Dutch masters. The potato was also an old host of viral diseases. In 1775, it was so severely attacked in certain parts of Europe that its cultivation in this instance had to be abandoned. Prizes were offered to those who could find the cause and remedy for the destructive ailment of such a useful plant.

Well before the development of bacteriology as a science, viruses were studied. In experiments on vaccination with cowpox, Edward Jenner, in the last quarter of the 18th century, used in crude form essentially the same general type of manipulation practiced in modern studies of viruses. The work of Pasteur on the development of a rabies vaccine in 1885 is well documented. The establishment of viruses as distinct pathological entities, however, did not come until the last half of the 19th century.

In 1857, Swieten described a mottling disease of tobacco leaves and Mayer named this condition tobacco mosaic, referring to the alternating patches of green- and brown-colored lesions on the leaves. Mayer showed the disease could be transmitted by injecting sap from affected plants into healthy ones. These transmission experiments can be considered as the starting point of modern studies on virus diseases.[1]

Iwanowski, in 1892, reported on the passage of diseased tobacco sap through a bacteria-proof filter, which was infectious when applied to healthy tobacco plants. However, he was certain that tobacco mosaic was a bacterial disease, and argued that the symptoms were produced by bacteria penetrating the filter, or by a toxin secreted by these organisms. Beijerinck (Figure 1.1) in 1898 conducted similar experiments and was concerned with the disease-causing filtrate, certain that the filter had held back bacteria. Beijerinck, convinced of the agent's filterability, was further impressed by its diffusion through agar gels. He considered the disease agent different from a bacterium. In his view, the causal microbe of tobacco mosaic was a *new* infectious agent, not corpuscular, but a fluid, a "contagium vivum fluidum."[2] This term may be loosely translated as a living germ that is soluble.[3]

While this description was puzzling — citing the agent as noncorpuscular and fluid, it did arouse numerous investigations in this new area of infectious diseases. The agent referred to by Beijerinck, the tobacco mosaic virus, would be the first submicroscopic pathogen to be identified. In 1898, Löffler and Frosch found that the agent of foot-and-mouth disease of cattle was also capable of passing a filter impermeable to bacteria.[4] This result made it clear that filterable agents were not solely responsible for disease in plants, but could have other hosts.

The first disease of humans shown to be caused by a filterable agent was yellow fever. From investigations sponsored by the U.S. Army Yellow Fever Commission in 1900–1901, Walter Reed (Figure 1.2) and colleagues proved the theory of Finlay, a Cuban physician and epidemiologist, that the agent was

FIGURE 1.1   Martinus W. Beijerinck (1851–1931). His transmission experiments on tobacco mosaic disease led to the establishment of viruses as distinct infectious agents. (By permission of the Kluyver Laboratory for Biotechnology at Delft University of Technology.)

FIGURE 1.2   Walter Reed (1851–1902). In 1899, he headed a commission that proved yellow fever was transmitted by a mosquito. More importantly, by getting rid of the mosquito, the disease was abolished. In 1901, he proved the causative agent was a filterable substance of the type discovered by Beijerinck. (Courtesy: The National Library of Medicine.)

FIGURE 1.3   Larva of Gypsy moth killed by a virus. (Courtesy: The U.S. Department of Agriculture — Forest Service.)

transmitted through mosquitoes. In 1928, a Rockefeller Foundation team would prove the microbe was a virus.[5]

In 1907 von Prowazek, using filtered material from diseased insects, was able to infect healthy silkworms. Bergold, in 1947, showed that this infection was caused by rod-shaped particles now known as a class of insect viruses, the nuclear polyhedral viruses[6] (Figure 1.3).

Yet another kindom was found to harbor filterable agents. Twort[7] and d'Herelle[8] in the period 1915–1917 discovered the phenomenon of "bacteriophagia," now known to be caused by the activity of viruses parasitic to bacteria. By the 1930s it was becoming clear that a wide variety of infectious agents would have to be included in the filter-passing group.

The term "virus" had an Old World connotation with "poison," and during the nineteenth century it became a synonym for microbe, Pasteur's term for an infectious agent. The modifier filterable was used with "virus" to denote its difference from bacteria. In this connection, Rivers, in 1928, published a compendium, *Filterable Viruses*. Gradually, the adjective filterable was dropped when referring to viruses as a new kind of infectious agent. One reason for this action was that some viruses did not easily pass through standard filter candles. Among microbiologists the term "virus" began to be used universally without qualification.[5]

After 1930 there was a steady development of physical and biological techniques for the study of viruses. A notable advance was made by Elford's use of "graded" filter membranes of known pore size, first developed by Bechold in 1907. This allowed for the first time an opportunity of assessing the actual size of the particles causing viral diseases. It also served to replace the vague idea of a "contagium vivum fluidum" with something more precise.[1]

A number of technical achievements were introduced to study viruses. In Germany, in the 1930s, Ruska and Knoll and coworkers constructed the transmission electron microscope.[9] This group produced photographs of particles of tobacco mosaic virus (TMV), bacteriophages, and poxviruses. The electron microscope became an indispensable instrument for the study of viruses (Figure 1.4).

Of equal importance was the research of Goodpasture, who showed that the virus of fowl pox could be grown in the tissues of the developing chick embryo. In 1949, Enders and colleagues used tissue culture methods for the growth of poliomyelitis virus. This work led to the immunization against polio[5] and also offered a method by which numerous new types of viruses could be isolated (Figure 1.5).[10]

Viruses were found to violate Koch's postulates for microbes. Thus, some viruses, notably the herpes virus types, were observed to remain latent in host cells for years. So too, bacteriophages exhibited lysogeny, the process of inclusion of the page nucleic acid within that of the host cell followed by dormancy.[11] Viruses were also observed to mutate readily, as in the case of the influenza virus. Today the human immunodeficiency virus (HIV), the agent of acquired immunodeficiency disease syndrome

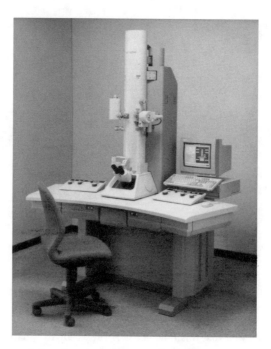

FIGURE 1.4 A modern transmission electron microscope, Model H-7500 TEM. (Courtesy: Hitachi Scientific Instruments.)

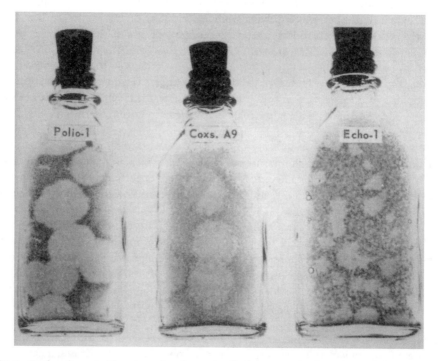

FIGURE 1.5 Bottle cultures in Rhesus monkey tissue of poliovirus type 1 (Mahoney), Coxsackie virus A9 (Grigg), and ECHO virus type 1 (Farouk), showing their characteristic plaque morphologies. (From Melnick, J.L., in *Cellular Biology, Nucleic Acids, and Viruses,* Special Publications of the New York Academy of Sciences, Vol. 5, 1957, 367. With permission.)

FIGURE 1.6 Max Delbrück (standing) and Salvador Luria at the Cold Spring Harbor Laboratory, 1941. (Courtesy: Cold Spring Harbor Laboratory Archives.)

(AIDS), is recognized as a highly mutable virus. Mutation results in a changed protein coat for such viruses, and in turn a changed virus immunologically.

Viruses have played a major role in understanding the flow of genetic information and biotechnology. Beginning with the establishment of the Phage School by Delbrück and Luria in the 1940s (Figure 1.6), at the Cold Spring Harbor Laboratory,[12] studies on bacterial viruses have played a seminal role in uniting the disciplines of biochemistry and genetics into molecular biology. In 1952 Zinder and Lederberg discovered transduction, the process whereby phages mediate bacterial recombination.[13] Berg and coworkers, in 1973, demonstrated the recombination of the DNA of bacteriophage lambda with the animal virus, Simian virus 40 (SV40).[5]

Messenger RNA (mRNA) was discovered as a result of phage studies. From 1956–1958, Volkin and colleagues studied RNA produced immediately after bacteriophage infection of *Escherichia coli.* Analyzing the RNA after incorporation of ^{32}P, they found that its base composition closely resembled phage DNA but was different from that of bacterial DNA. While the newly synthesized RNA was short lived, its production was shown to precede the synthesis of RNA as a preliminary step in the process of protein synthesis.[14]

In 1961, Brenner, Jacob, and Meselson extended these studies.[15] Uninfected *E. coli* ribosomes were labeled with heavy isotopes and then phage infection allowed to proceed in the presence of radioactive RNA precursors. It was demonstrated that phage proteins were synthesized on bacterial ribosomes that were present prior to infection. The ribosomes appeared to be nonspecific, strengthening the possibility that another type of RNA serves as an intermediary in protein synthesis. At this time, Spiegelman and coworkers isolated ^{32}P-labeled phage RNA following infection of bacteria and studied its hybridization potential with phage DNA. The RNA was complementary in base sequence to the viral genetic information.[16] The results of these diverse studies agree with the concept of a messenger RNA, mRNA, being

FIGURE 1.7   Martha Chase and Alfred Hershey at the Cold Spring Harbor Laboratory (early 1950s). (Courtesy: Cold Spring Harbor Laboratory Archives.)

produced on a DNA template and then directing the synthesis of specific proteins in association with ribosomes.[17]

The ability to manipulate viruses for the expression of genes was realized with the development of molecular cloning techniques in the 1970s. Initially, SV40 and bovine papilloma virus were utilized to vector foreign genes into mammalian cells. The first genetically engineered human vaccine, yeast-grown hepatitis B vaccine, was licensed in 1986.[5] The first human clinical trials for an AIDS vaccine used an HIV gene product purified from a baculovirus expression system.[18] At present, virtually all types of viruses may be used as vectors of genes from heterologous sources.

The structural composition of viruses consists of two major components: nucleic acid, surrounded and protected by protein. The nucleic acid component carries the genetic instructions to which a host cell submits in producing more virus particles. The nucleic acid, either deoxyribonucleic acid or ribonucleic acid (DNA and RNA, respectively), may be double or single stranded. Hershey and Chase (Figure 1.7), in experiments involving radioactive labeling of sulfur in the protein component and of phosphorus in the nucleic acid component, noted that after attachment, viral multiplication depended on the nucleic acid entity.[19] Later, Gierer and Schramm[20] and Fraenkel-Conrat and coworkers[21,22] demonstrated the RNA component of tobacco mosaic virus as the infectious entity. These experiments pointed to nucleic acid as the heredity molecule and complemented the work of Avery, McCleod, and McCarty, in which nucleic acid was shown to be the transforming factor of pneumococcus bacteria.[23]

Depending on the type of nucleic acid, viruses may reproduce in a variety of ways. Three common pathways will be briefly discussed. The first pathway follows the semiconservative replication of DNA, which leads to a messenger RNA and then to protein production. The T even phages are examples. This synthesis is called the central dogma.[24]

Viruses possessing RNA follow one of two pathways. The RNA may act as the single stranded carrier of genetic information; no intervening DNA is utilized, and protein production results from the genetic direction of RNA. Polio virus and TMV are examples. The other RNA virus pathway for reproduction involves reversal of the common pathway. RNA is first converted to a single stranded DNA, from which proteins are assembled. This reversal of roles is found especially in retroviruses, e.g., the Rous sarcoma virus and HIV. The enzyme for converting RNA to DNA, reverse transcriptase, was discovered independently by

Howard Temin[25] and David Baltimore.[26] Reverse transcriptase has two major enzymatic functions: a DNA polymerase by which DNA is formed from the RNA template, and an RNA function. The latter function degrades the original RNA once the DNA is formed.

Reverse transcriptase produces DNA complementary to mRNA. In this transcription, the segments of DNA not directly involved in the flow of genetic information, introns, are omitted. The segments retained, exons, then represent a fraction of the original DNA. This feature has been used in the Human Genome Project, a study whereby genes and their functions are to be identified in chromosomal DNA. J. Craig Venter has employed reverse transcriptase to obtain complementary DNA, cDNA, from messenger RNA.[27] This act greatly facilitates the identification of nucleotides and was used to determine the complete coding of genes in the DNA of the bacterium, *Hemophilus influenzae*.

The fact that RNA can, like DNA, carry genetic information has given credence to the concept of an RNA world in the beginning of life. The primacy of DNA over proteins and RNA has been shaken by the long realization but underemphasis of the existence of RNA in viruses requiring no apparent DNA role in multiplication. Another positive consideration for RNA as a key molecule in evolution is the capability to act enzymatically. Thus RNA is capable of exercising two principle activities required in living forms, genetic passage and enzyme action.

While the existence of viruses as infectious entities has been known for a century, viral diseases appear to have originated with the beginning of living forms. Hybridization analysis has shown that there are virus genes derived from baboon endogenous retroviruses in all Old World monkey species, and that they have co-evolved with their hosts. Such viruses, it is believed, are at least 35 million years old.[28]

Viruses have been used to study evolution on the molecular level. The Qβ bacteriophage, a small virus that infects the bacterium *E. coli*, contains only four genes, one of which encodes a protein enzyme called a replicase. The replicase is vital to the survival and proliferation of the phage because it makes copies of the viral RNA genome. In 1965, Spiegelman and coworkers demonstrated *in vitro* replication of infectious viral Qβ RNA. Viral RNA was mixed with its replicase protein and ribonucleoside triphosphates. The RNA replicated *in vitro* and when added to *E. coli* protoplasts, infection and viral multiplication resulted. In these experiments RNA synthesized in a test tube served as genetic material.[29]

The Qβ RNA replicase system has the property of amplification, as demonstrated by Spiegelman et al. In addition, the process has a built-in mutation feature since the viral replicase always makes one or two errors when it copies a sequence of RNA. As a selection criterion, time was chosen as the governing factor and a serial transfer experiment was conducted by Spiegelman. He allowed the Qβ replicase to amplify Qβ RNA in a test tube for 20 minutes. During this period, the replicase made many copies of the parent molecules as well as copies of the progeny, committing occasional errors in the process. Spiegelman then transferred a sample of the reaction mixture to another test tube containing a fresh supply of replicase enzyme and nucleoside triphosphates. The cycle of amplification and sample transfer was repeated 74 times. In each cycle the procedure favored the proliferation of those molecules that gave rise to the most progeny before the transfer step.[29]

If the time available for amplification was shortened, selection would favor those molecules that could be produced during this time. By the 74th transfer the evolving RNA molecules had lost 83% of the original Qβ genome, retaining only that part the replicase needed to exercise its function. The Qβ system served as a prototype for similar types of molecular evolution experiment.[30,31]

Viruses have been known to induce tumors and cancers in lower forms since their earliest observations. In 1911, Rous (Figure 1.8) reported on a transmissible agent causing sarcoma of chickens.[32] Later this effect, attributable to a retrovirus, the Rous sarcoma virus, was reported in other species: murine leukemia viruses and avian leukosis viruses.[33]

Viruses capable of "inducing" tumors are diverse, and include RNA and DNA viruses.[34] The retroviruses have contributed much to our understanding of virus-induced tumors. The retroviral action typically follows the nucleic acid pathway common to this type of virus. The enzyme reverse transcriptase converts the RNA of the virus to a complementary DNA, which is integrated into the DNA of the host cell, and in this situation may produce tumors in weeks or over longer periods of time.

FIGURE 1.8   Francis Peyton Rous (1879–1970). In 1911, he reported a chicken sarcoma caused by a filterable agent. Later, the agent would be recognized as the first of the tumor viruses. (Courtesy: The Rockefeller Archive Center.)

Some human cancers have a viral etiology. Burkitt's lymphoma appears to have a relationship with the DNA-containing Epstein-Barr virus. Human cervical cancers are strongly associated with DNA-containing papilloma viruses. The causative agent of human T-cell leukemia is a retrovirus. The relatively recent successes with organ transplantation is clouded by the possibility of infection by microbes, including viruses, residing in the donor's transplanted organ. Some of these viruses may be cancerous.

With the discovery of large numbers of different viruses, a system was needed for their classification. In 1966 the International Committee on Nomenclature of Viruses was established at the International Congress of Microbiology in Moscow. In 1973 the name was changed to the International Committee on Taxonomy of Viruses (ICTV). Presently, the ICTV operates under the auspices of the International Union of Microbiological Societies. The ICTV has six subcommittees, 45 study groups, and its duties are aided by over 400 participating virologists.

The present universal system of virus taxonomy is "useful and usable".[35] It is set arbitrarily at hierarchical levels of order, family, subfamily, genus, and species. Virus orders represent groupings of families of viruses that share common characteristics and are distinct from other orders and families. Virus orders are designated by names with the suffix *-virales*. Virus families represent groupings of genera of viruses that share common characteristics and are distinct from the member viruses of other families. Virus families are designated by names with the suffix *-viridae*, and the subfamily ending is *-virinae*. Virus genera represent groupings of species of viruses that share common characteristics and are distinct from the member viruses of other genera. Virus genera are designated with the suffix *-virus*. The species taxon, while the most important hierarchical level in classification for viruses, is the most difficult to clearly define. The ICTV study groups are presently determining specific properties to be used to define species in the taxon for which they are responsible. In the meantime, some viruses have already been designated as species, e.g., Sindbis virus, Newcastle disease virus, poliovirus I.[35]

In formal taxonomic usage, the first letters of virus order, family, subfamily, and genus names are capitalized and the terms are printed in italics, or underlined when written. Species designations are not capitalized, unless they are derived from a place name or a host family or genus name, nor are they italicized. In formal usage, the name of the taxon should precede the term for the taxonomic unit; for example, the family *Paramyxoviridae*.

FIGURE 1.9   Theodore Diener (1989), discoverer of viroids. (Courtesy: Dr. T. O. Diener.)

The following represents an example of a full taxonomic terminology: order, *Mononegavirales;* family, *Rhabdoviridae;* genus, *Lyssavirus;* species, rabies virus.

The Sixth Report of the ICTV, published in 1995, records a universal taxonomy scheme comprising one order (*Mononegavirales*), 71 families, 9 subfamilies, and 164 genera, including 24 floating genera, and more than 3600 virus species. The system still contains hundreds of unassigned viruses, largely because of a lack of data.[35]

Virology presents a continuing updating process for taxonomists. Some rod-shaped viruses exhibit isolated functions. In tobacco rattle virus,[36,37] the RNA replicative function exists in some particles and the RNA coat function in others. Some viruses require a helper virus. Tobacco necrosis virus particles include smaller particles, satellite viruses, which require the assistance of normal particles to carry out replicative functions.[38] Some infectious entities are tentatively classified as subviral, because they appear to lack either the nucleic acid or protein component usually found together in viruses. In this connection, Diener (Figure 1.9), in 1971, discovered small infectious RNA strands without protein.[39] These RNA strands, termed "viroids" by Diener, were able to produce a disease of potato spindle tuber. A number of viroids have been isolated and studied. While all have been assigned a plant host function, it is believed that this type of infectious entity will be discovered in higher forms as well.

Perhaps, the most baffling of subviruses are prions[40] — infectious proteins having no apparent detectable nucleic acid involvement. This observation runs counter to the nearly universal concept that nucleic acids are the infectious components of disease agents. Extensively investigated by Gadjusek at the U.S. National Institutes of Health,[41] and later by Prusiner at the University of San Francisco, prions are suspect as the putative disease agents of scrapie, kuru, and Creutzfeld-Jacob diseases.

1.1   The Attraction of Physicists and Chemists to Biology

A number of events along with considerations of applying physical laws to biological processes attracted scientists to biology. One consideration was an understanding of the concept of the biological macromolecule. Macromolecules are a reality, a concept championed by Staudinger, and discussed in Chapter 2. An important event was the report of the crystallization of tobacco mosaic virus in 1935 by Wendell Stanley (Figure 1.10). This communication was hailed by the scientific community. Especially, it caught

FIGURE 1.10   Wendell M. Stanley (1904–1971). His early studies on viruses led to an increased interest in the emerging discipline of virology. (Courtesy: The Rockefeller Archive Center.)

the interest of scientists like Hermann J. Muller, the first to use X-rays to achieve artificial mutations in a living form, the fruit fly, *Drosophila melanogaster*.

Muller linked viruses to genes and called for nonbiological scientists, especially physicists, to take up where Stanley had shown the way — an investigation of genetics as the science intimately concerned with life.[42] Max Delbrück began his professional life as a physicist; however, he was led into virology through intimate discussions with the quantum physicist Niels Bohr. Bohr advocated the application of physical concepts to biology. Bohr's influence may be summed up in a lecture delivered before the International Congress of Light Therapy in 1939. His address titled, "Light and Life,"[43] had as its theme that physical and biological investigations are not directly comparable. Delbrück adopted this view in his approach to biological study — that genetics was, in fact, a domain of biological inquiry in which "normal" physical and chemical explanations might turn out to be "insufficient." His philosophy of the need for "other laws" was stated in an address in 1949, titled "A Physicist Looks at Biology."[44]

Stanley's work interested Delbrück and, in surveying the field of virology for a model to study biology on a "molecular" level, he settled on bacteriophages. "I was absolutely overwhelmed that there were such very simple procedures with which you could visualize particles. I mean, you could put them on a plate with a lawn of bacteria, and the next morning every virus particle would have eaten a macroscopic one-millimeter hole in the lawn. You could hold up the plate and count the plaques. This seemed to be just beyond my wildest dreams of doing simple experiments on something like atoms in biology."[45]

The Phage School at Cold Spring Harbor attracted scientists of varied interests. For example, in 1947, Leo Szilard, the atomic physicist, took the phage course at Cold Spring Harbor.[46]

Delbrück's practical but nonclassical interest in biology arose from a collaboration with the Russian geneticist, Timoféeff-Ressovsky, and with Zimmer, a radiologist. The three published a study on the X-ray mutagenesis of *Drosophila*, "On the nature of gene mutation and gene structure."[47] Delbrück attempted to define the results in terms of quantum mechanics. His contribution "Physical-atomic Model of Gene Mutation," became known as a quantum mechanical model of the gene. An estimate for the minimum size of the gene was actually given. Professor Olby, in his valuable book, *The Path to the Double Helix*,

FIGURE 1.11 Erwin Schrödinger (1887–1961), the author of *What is Life?*, induced many scientists to pursue studies of a biological nature. (Courtesy: American Institute of Physics Emilio Segrè Visual Archives, Francis Simon Collection.)

The Discovery of DNA, called this paper "… of fundamental importance in the history of molecular biology."[48]

Erwin Schrödinger (Figure 1.11), in his famous book, *What is Life?*, emphasized Delbrück's physico-mathematical approach to genetics.[49] Schrödinger's book attracted scientists who would concentrate on molecular aspects of biology: Luria, Crick, Wilkins, Benzer, and Watson. Schrödinger's central theme was that a codescript for living things exists. The codescript is contained in an aperiodic crystal — a crystal that is not a repeat of the same unit, but varies in its makeup. George Gamow, a physicist famous for his Big Bang theory of the universe, suggested that nucleic acids acted as a "genetic code" in the formation of enzymes. He was the first to maintain that the code was made up of triplets of nucleotides.[50]

Investigators concerned with the flow of genetic information were part of what Olby refers to as the informational school. There was another line of inquiry, one concerned with the structure of macromol-ecules such as proteins and nucleic acids — in Olby's words, the structural school. The intellectual breeding grounds for this group of investigators was Europe and began with the use of X-rays to study giant molecules.[48]

Olby notes how the structural schools of Leeds, Cambridge, and Caltech and the informational school, born at Cold Spring Harbor, were united in the team of Crick and Watson. Crick could be judged to be a member of the structural school in his pursuit of X-ray studies. Watson, initially, was a member of the informational school, with his genetic leanings fostered by Luria and Delbrück. In their union, Crick and Watson, in large part, paved the way in solving the secret of genetic information through structural studies on DNA.

Olby also notes the importance of the Rockefeller Foundation whose support influenced the migration of scientists into the study of biology. From 1932–1959, the Foundation spent over $90,000,000, a large part of it on biological research supported by its president, Max Mason (Figure 1.12). Thus, investigators such as Muller, Delbrück, Caspersson, Hammarsten, Astbury, Pauling, Corey, and Svedberg, all received support from the Foundation during this period. Out of the Rockefeller Foundation's interest in the new

FIGURE 1.12   Max Mason (1877–1961). As Rockefeller Foundation president, he supported research in the "new Biology." With Warren Weaver, he is considered a pioneer in sedimentation analysis. (Courtesy: The Rockefeller Archive Center.)

FIGURE 1.13   Warren Weaver (1894–1978). As director of the natural sciences division of the Rockefeller Foundation, he actively supported research in developing areas of biology. He coined the term "molecular biology." "Weaver's rule" dealt with the time required to attain equilibrium in a gravitational field. (Courtesy: The Rockefeller Archive Center.)

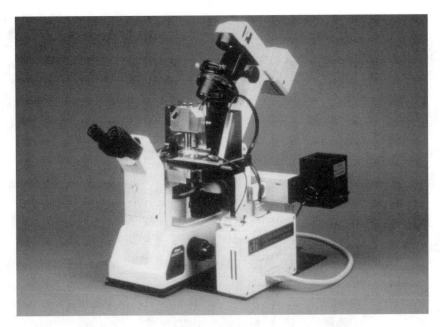

FIGURE 1.14 BioScope™ Atomic Force Microscope. (Courtesy: Digital Instruments, Santa Barbara, CA.)

biology came the name best suited to describe it — molecular biology. In 1938, the director of the National Sciences section of the Foundation, Warren Weaver (Figure 1.13), used the term to describe the Foundation's commitment to this branch of science:[51]

And gradually there is coming into being a new branch of science — molecular biology — which is beginning to uncover many secrets concerning the ultimate units of the living cell…. Among the studies to which the Foundation is giving support is a series in a relatively new field, which may be called molecular biology in which delicate modern techniques are being used to investigate ever more minute details of certain life processes.

The term was brought into prominence with the establishment by Perutz of the Laboratory of Molecular Biology at Cambridge University in the 1940s. In 1959, the *Journal of Molecular Biology* was introduced to the scientific community, with the objective of elucidating this new discipline of science.

In the study of viruses new methodologies are employed almost as soon as they emerge. One of the most exciting new instrumentation technologies is the atomic force microscope (AFM)[52] (Figure 1.14). The AFM belongs to a class of scanning probe microscopes, instruments capable of generating nondestructive, three-dimensional surface profiles of molecular structures with nanometer resolution.

The AFM, an offshoot of the scanning tunneling microscope,[53] provides detailed topographic maps of sample surfaces by raster scanning a fine tip gently over the surface of a specimen. Atomic force microscopy is being applied to virological research. Single living cells infected with viruses have been observed with the AFM,[54] and it has been employed to image viruses[55] (Figure 1.15) and viral nucleic acid[56] (Figure 1.16).

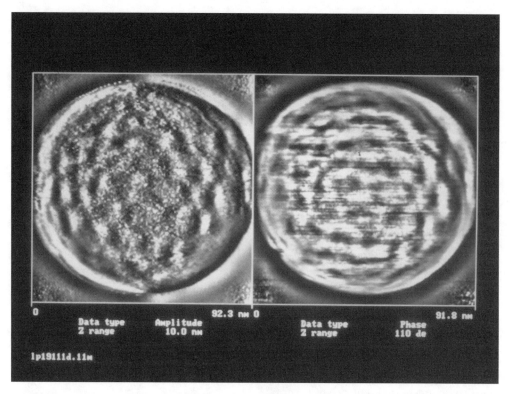

FIGURE 1.15   Simultaneously acquired amplitude (left) and phase (right) images of a single rotavirus particle using Tapping Mode in air. (Courtesy: Digital Instruments, Santa Barbara, CA.)

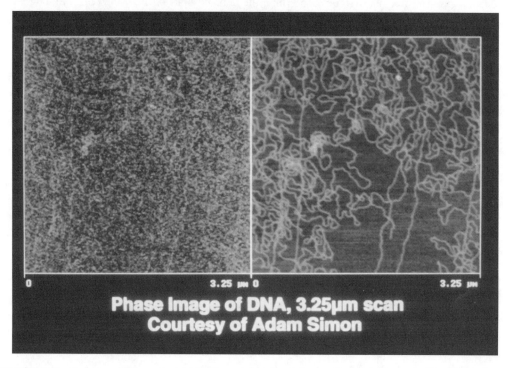

FIGURE 1.16   Simultaneously acquired (a) height and (b) phase images of genomic DNA from the lambda bacteriophage scanned in air on a mica substrate. 3.25 μm scan. (Courtesy of Digital Instruments, Santa Barbara, CA.)

2

Viruses as
Macromolecules

2.1 The Macromolecule Concept

"For such colloidal particles in which the molecule is identical with the primary particles, where the individual atoms of the colloidal molecule are those bounded by normal valency activity, we suggest the term *Macromolecule*".[57]

"One of the most interesting aspects of protein research, and one which has only recently emerged, is the indication that these huge molecules exhibit phenomena that we ordinarily consider possible only to living organisms. Thus viruses 'reproduce', when in a suitable environment; and yet the brilliant researches of Stanley and others have shown that certain viruses which show this property so characteristic of life are nothing more than huge protein molecules."[58]

"While no harm is done by calling viruses 'molecules', such terminology should not prejudice our view regarding the biological status of viruses, which has yet to be elucidated."[59]

That viruses are macromolecules arises from a consideration of their essential components, protein and nucleic acid. Proteins and nucleic acids are recognized as macromolecules, but that designation was the result of a tortuous and sometimes highly debated course.

The idea of naturally existing, very large molecules arose in the 19th century, and the element carbon was central to such a concept. Of prehistoric discovery, the element carbon has a principal valence of four. This property allows carbon atoms to bond covalently with other atoms to form a stable system of eight electrons, i.e., an electron octet. The German chemist Kekulé (Figure 2.1) in 1858 suggested that carbon was tetravalent and proposed this fact as a law for carbon compounds.[60] In addition, he maintained that one, two, or three of the four bonds of a carbon atom could attach to another carbon atom so that chains or polymers of such atoms could be formed — "Kekulé structures."[61] Kekulé's concept of a chain of carbon atoms attached to other atoms was based on the combining power of atoms, referred to as valence (from the Latin word for power). In 1852, Edward Frankland (Figure 2.2), an English chemist, pointed out that each kind of atom can combine with a limited number of other atoms: hydrogen, with a valence of one can only combine with one other atom; oxygen, with a valence of two, can combine with a maximum of two other atoms. Carbon, having a valence of four, is able to form long chains of molecules.[62] The theory of valence dominated the subsequent development of structural doctrine and formed the basis of modern structural chemistry.

Another important concept concerning structural chemistry was the three-dimensional form of molecules. This idea was put forth independently in 1874 by van't Hoff (1852–1911) in Holland, and Le Bel (1847–1930) in France (Figures 2.3 and 2.4).

Molecules containing carbon were viewed as three-dimensional objects instead of a two-dimensional arrangement. The carbon atom served as the center of a pyramid, with its four bonding arms radiating out to the points of the pyramid. With this three-dimensional view, multiple bonding could also be explained. Single bonds between carbon atoms could be accounted for by placing the pyramids point-to-point; double bonds, by placing the carbon atoms edge-to-edge; and triple bonds, by placing the carbon atoms face-to-face[62] (Figure 2.5).

FIGURE 2.1   Friedrich August Kekulé (1829–1896). (Courtesy: The Chemical Heritage Foundation.)

FIGURE 2.2   Sir Edward Frankland (1825–1899). (Courtesy: Department of Special Collections, Van Pelt-Dietrich Library Center, The University of Pennsylvania.)

FIGURE 2.3 Jacobus H. van't Hoff (1852–1911). (Courtesy: The Chemical Heritage Foundation.)

FIGURE 2.4 Joseph A. LeBel (1847–1930). (Courtesy: Department of Special Collections, Van Pelt-Dietrich Library Center, The University of Pennsylvania.)

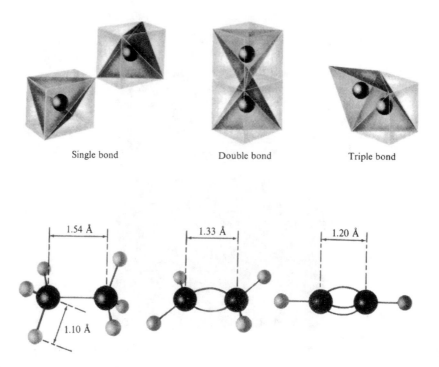

FIGURE 2.5   Top: Tetrahedral atoms forming single, double, and triple bonds. Bottom: Valence-bond models of ethane (single bond), ethylene (double bond), and acetylene (triple bond). (From Pauling, L., *General Chemistry*, Dover Publications, New York, 1988. With permission.)

The diversity and complexity of organic forms result from the capacity of carbon atoms for uniting with each other in various chain and ring structures and three-dimensional conformations, as well as the tendency for linking with other atoms. Clearly organic chemistry is the chemistry of carbon compounds, Kekulé's definition, as given in his textbook, *Lehrbuch der Chemie* (1859).[63]

The ability to make large molecules out of smaller ones was demonstrated by Fischer (Figure 2.6) in his work on proteins and carbohydrates. It was known that large molecules were built up of monomers; amino acids in the case of proteins. Fischer showed exactly how these amino acids were combined with each other within the protein molecule through the formation of the peptide bond. In 1907, he synthesized a simple but authentic protein molecule made up of 18 amino acid units, 15 of glycine and 3 of leucine, and showed that digestive enzymes attacked it just as they would attack natural proteins (Fischer, 1914).[64,65] This experiment was a significant beginning in the complex field of protein structure.

The prevailing theory concerning giant molecules in the first decades of the twentieth century, however, was that they were *aggregates* of smaller molecules. Such aggregates possessed no true chemical bonding, but were held together by weak, physical, intermolecular forces. This view was endorsed by Fischer, who denied the existence of or necessity for giant polypeptides.[65]

The aggregate theory for explaining the presence of giant molecules was based on three ideas, as noted by Olby.[48] First, the cofounder of coordination theory, Alfred Werner (Figure 2.7), introduced the concept of two kinds of combining forces in chemical compounds — primary valence forces (hauptvalenzen), and secondary valence forces (nebenvalenzen).[66] According to Werner, atoms united by primary valence forces, e.g., covalent bonds, possessed varying degrees of "residual affinity," whereby simple molecules could become united into "compound molecules" or aggregate molecules. The residual affinity was the result of secondary valence forces. This idea, cites Olby, was applied by Karrer and Hess to starch and cellulose, by Plummerer and Harries to rubber, by Bergmann to proteins, and by Hammarsten to thymonucleic acid, the initial name for DNA.

FIGURE 2.6   Emil H. Fischer (1852–1919). (Courtesy: The Chemical Heritage Foundation.)

FIGURE 2.7   Alfred Werner (1866–1919). (Courtesy: Department of Special Collections, Van Pelt-Dietrich Library Center, The University of Pennsylvania.)

Second, in analyzing the unit cells of giant molecules some X-ray crystallographers concluded that such molecules could not be larger than the unit cell. Since the unit cell was small, so too, must be the molecules.

The third basis for support of the aggregate theory of giant molecules was the concept of colloidal matter. In this connection, Professor Flory in his book, *Principles of Polymer Chemistry,* notes the influence of Thomas Graham. Graham, one of the founders of physical chemistry, known for his relationship that the rate of diffusion of a gas is inversely proportional to the square root of its molecular weight, called attention to the slow or negligible rates of diffusion of certain polymers in solution. Such substances were unable to pass through parchment, a semipermeable membrane, and were termed by Graham "colloids" (Greek *kolla*, glue). For substances that did pass through the parchment, especially those that can be obtained in the crystalline state, he used the term "crystalloids." It was believed that many substances could aggregate to become colloidal in size. Colloidal chemists such as Ostwald advanced the idea of the colloidal state of matter as a physical rather than a chemical state of organization, in which numerous molecules of ordinary size were held together by intermolecular, "secondary valence" forces — physical attractions.[67]

The idea of the colloidal state of matter was widely regarded in science. *Protoplasm,* a term for the colloidal embryonic material in the egg, was coined by Purkinje, a Czech physiologist, in 1839. In 1846, von Mohl, a German botanist, used the term to distinguish between the watery, nonliving sap in the center of a cell and the granular, colloidal living material rimming the cell. Eventually, the word came to mean the living material within a cell. Subsequently, the English biologist Huxley referred to protoplasm as *"the physical basis of life."* Cohn, a German botanist, showed that in plant and animal cells, the "protoplasms" were essentially identical, and that there was, therefore, only one physical basis of life. Colloid science, not strictly obeying the common laws of chemistry, appeared to be well suited to clarifying the concept of this living substance.[68]

The term colloid was also applied, erroneously, to other forms of matter having little in common with macromolecular substances. Thus, gold sols, soap solutions, and colloidal solutions of tannic acid resembled solutions of high polymers. This misuse of the meaning of colloids was referred to by Florkin in his book, *A History of Biochemistry,* as the "age of micellar biology."[69]

However, a small number of investigators held the view that some natural substances such as cellulose, starch, rubber, proteins, and carbohydrates were polymers. Flory notes that the idea that proteins and carbohydrates are polymeric goes back at least to Hlasiwetz and Habermann, who in 1871, considered that these substances included a number of isomeric and polymeric species differing from one another with respect to the degree of molecular condensation.[67] And Kekulé had noted that carbon atoms could link to form "netlike" and "sponge-like" molecular masses which resisted diffusion.[60]

Many investigators using reliable methods of the time to determine molecular weights: osmotic pressure, freezing point depression, boiling point elevation, failed to read in their results that polymers such as cellulose and starch were composed of very large molecules. They questioned the applicability of physicochemical laws to colloids. In such cases, where negligible freezing point depressions and boiling point elevations were noted in solutions where the colloid was dispersed, this behavior was considered "normal." As we know, these colligative methods for molecular weight determinations are based on the *number* of solute particles in solution and not on the *weight* of the molecules.[67]

Flory offers a cogent argument in differentiating between aggregated masses of colloidal molecules and true polymers:

> …it is conceivable that any molecular substance may, under suitable conditions, aggregate to particles which are nevertheless small, i.e., of colloidal dimensions. The implied converse of the concept of colloid particles as the manifestation of a state of matter does not follow, however. That is to say, many colloidal substances (as originally defined by Graham) are known which do not revert to "crystalloids" without chemical change. Thus, the individual molecules of cellulose and or high molecular weight polystyrene are typical colloids according to the intent of Graham's definition, but they cannot be disaggregated by any process corresponding to a physical change of state. A "crystalloidal solution" of such substances is therefore unattainable. Hence, the concept of the colloidal state as a purely physical state of organization is inapplicable to the very substances for which the term colloid was originally chosen. For many years investigators were seldom concerned with, or aware of, the distinction between a colloidal particle composed of numerous molecules of ordinary size held together by intermolecular "secondary valence" forces of one sort or another and a polymer molecule made up of atoms held together exclusively by covalent bonds.[67]

FIGURE 2.8   Hermann Staudinger (1881–1965). (Courtesy: The Chemical Heritage Foundation.)

The period 1926 to 1930 was a time when meaningful challenges to the aggregation or secondary valence viewpoint took hold. The transition was not a smooth one. The chief proponent of the view that large natural and synthetic substances were constituted of primary valence (covalent) molecules was the German organic chemist, Staudinger (Figure 2.8). In a paper published in 1920, he deplored the prevailing tendency to formulate polymeric substances as association compounds held together by "partial valences." He held that synthesized polymers such as polystyrene and paraformaldehyde, as well as natural ones, such as rubber, were held together by primary valence bonds.[67] In 1924, Staudinger introduced the term macromolecule as a synonym for high polymer. Macromolecules were composed of monomers united by main (hauptvalenzen or covalent) valences. Many colloidal particles are aggregates of hundreds of molecules, but others, such as proteins and polymer molecules, consist of a single large molecule.[57]

Opposition to Staudinger's macromolecule theory was strong, but support of his views became firmly established by 1930. In this connection three approaches pointed to the existence of high polymers as valid, covalently bonded structures.[48]

First, the fallacy in the assumption that a molecule could be no larger than its unit cell was pointed out by Polyani in 1921, and by Herzog in 1926. The space lattice of a crystal may be considered as built up of a three-dimensional basic pattern, one that is continuously repeated. This repeating unit is called the unit cell and the external appearance of the crystal is determined by its shape and dimensions. Sponsler and Dore in 1926 demonstrated that the results of X-ray diffraction by cellulose fibers are consistent with a chain formula composed of an indefinitely large number of units.[70] The structural units occupy a role analogous to that of the molecule of a monomeric substance in its unit cell. The cellulose molecule continues from one unit cell to the next through the crystal lattice. Neither the dimension of the unit cell, nor those of the entire crystal, are related, directly, to the chain length of a polymer. The molecule of the polymer may pass through many unit cells reaching from one end of the crystal to the other, then pass through an amorphous region into another crystal, and so on.[67]

A second approach in supporting the molecular viewpoint concerning polymers was that of the synthesis of polymeric molecules of definitive structures. The work of Carothers at the Dupont Chemical Company serves as a notable example. In 1929, he set out to prepare polymers through the use of established reactions of organic chemistry, and then proceeded to investigate how the properties of these

FIGURE 2.9   Theodor Svedberg (1884–1971). (Courtesy: Beckman Instruments, Palo Alto, CA.)

substances were dependent upon their constitution. In this connection, Carothers and others extended the synthetic approach initiated by Fischer.[63]

The third consideration, one that was convincing to many that macromolecules were real substances, was the invention of the ultracentrifuge by Svedberg (Figure 2.9). In 1924, he completed the construction of the first "ultracentrifuge," and succeeded in determining the molecular weights of proteins, and later of viruses. Centrifuges had been used as a source of artificial gravity to separate light and heavy particles in a suspension, just as a farmer uses a simple centrifuge to separate light cream from heavy milk by spinning the liquid at a high speed. Centrifugal force throws heavier particles farther from the axis of rotation than lighter particles. Svedberg built a centrifuge with speeds not attained up to that time. By using a rotor driven by an oil turbine, he increased the speed of the centrifuge to some 60,000 revolutions per minute, with a centrifugal force of some 250 times the pull of gravity (Figure 2.10). With this high force Svedberg measured the settling, or sedimentation, rates of many molecules, and then calculated the molecular weights of the molecules. He found that proteins were indeed giant molecules, macromolecules, having weights greater than those of other "ordinary" molecules. Macromolecules were like smaller molecules, both having covalent bonds. The major difference between them was simply that of size.[71]

Viruses exhibit a bit of irony with regard to aggregates and true macromolecules. Composed of two types of macromolecules, proteins and nucleic acid, aggregation does occur, in the formation of the intact virus. Subunits of protein aggregate to form the structural arrangement of the protein coat in surrounding the inner nucleic acid. Also, weak forces do play a very important role in viral structural organization. These factors are discussed below.

2.2   Proteins

From their discovery in the 19th century, proteins were recognized to be of central importance in living organisms. The designation protein had been coined by chemists from the Greek word, *proteios,* meaning "holding first place." In 1897, Buchner extracted zymase from yeast cells. For the first time cell-free extracts of enzymes could be studied. It was known that enzymes were proteins, and Buchner's work gave new emphasis to the importance of proteins in living forms.[72]

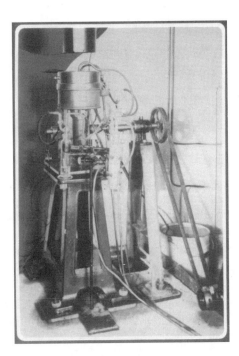

FIGURE 2.10   An early centrifuge used by Svedberg in his sedimentation studies. (Courtesy: Beckman Instruments, Palo Alto, CA.)

Notable academics of the late 19th and early decades of the 20th century championed the role of proteins. In America, the biologist Jacques Loeb held that proteins were the principal constituent in all matter and that their physicochemical properties governed all animate processes.[73] In 1916, the Harvard psychophysicist Troland, in his article "Biological Enigmas and the Theory of Enzyme Action," urged the view that the genetic "unit characters" were enzymes.[74] At about this time the term "autocatalysis" appeared in the literature. It described organismic and chemical growth as exponential functions. Thus enzymes, chromosomes, and genes expressed autocatalytic properties.[75]

The nature of viruses was deliberated after their isolation was validated. In this connection, TMV was proposed to be enzymatic by Woods in 1899–1900; the necrotic spots on infected leaves were due to a local accumulation of oxidizing enzymes.[76] The Dutch geneticist Hagedoorn likened genes to filterable viruses, and expressed the view that viruses were "chemical substances with auto-catalytic properties." Hagedoorn supported the concept of autocatalysis, and in reference to viruses, stated, "The discovery of the filterable viruses and bacteriophages thus offered valuable support to the enzyme theory of life and to the conception of autocatalysis as the fundamental and primitive characteristic of life."[77]

The discovery of bacteriophages led to a consideration that the agent of bacterial lysis was an enzyme produced by the bacterium. This idea was considered a possibility by Twort, a codiscoverer of phages.[78] The Nobel prize winning enzymologist Northrop, who later became interested in phages, proposed an intimate relationship between viruses, proteins, and autocatalysis. "The cells synthesize a 'normal' inactive protein. When the active virus or bacteriophage is added, this inactive protein or 'prophage' is transformed by an autocatalytic reaction into more active phage."[79]

Some laboratory experiments suggested that viruses were proteins. TMV could be precipitated using protein precipitants.[80] From a lead acetate precipitate of the virus Vinson, at the Boyce Thompson Institute for Plant Research in New York, determined nitrogen levels.[81] Dvorak in 1927[82] and Purdy in 1929[83] demonstrated the serological activity of TMV as evidence of its protein nature. In 1936 came the stunning announcement from Stanley: "Isolation of a crystalline Protein Possessing the Properties of TMV." Stanley stated he had "strong evidence that the crystalline protein herein described is either pure or is a solid solution of an autocatalytic protein which for the present may be assumed to require the presence of living cells for multiplication."[84]

Professor Lily Kay, in her book, *The Molecular Vision of Life,*[85] noted the euphoria that Stanley's report created among scientists and in the media:

> Life scientists and the popular media alike credited Stanley with finding the key to the riddle of life. Muller touted Stanley's work as an epochal discovery, spreading the lessons of protein crystals among geneticists. George Beadle, Max Delbruck, and the officers of the Rockefeller Foundation singled out Stanley's work as the most important breakthrough in understanding the molecular basis of the gene: the discovery has been described as the symbolic beginning of molecular biology.[86]

Then from England came the report by Bawden of the Rothamsted Experiment Station and Pirie of the Biochemistry Department at Cambridge University, that TMV was a nucleoprotein.[87] The path of study was now laid out. Viruses had a protein component and a nucleic acid component.

What was the nature of proteins? Were they mixtures or aggregates? Could a definite size be ascribed to them? In retrospect, Professor Flory notes:

> Emil Fischer considered his eighteen-membered synthesized polypeptide to be similar in molecular weight to most natural proteins, although he recognized that different amino acids were involved and that their order along the chain did not correspond to that occurring in proteins. His polypeptide theory is accepted today, but his estimate of the length of the chain was too small by one-to-four orders of magnitude, depending on the protein.[67]

Initially, following the enormous success of Fischer in working out the three-dimensional structures of proteins, it was supposed that such molecules consisted of long chains of repeated peptide units *aggregated* together and bonded by weak intermolecular forces. However, as noted by Professor Edsall, other workers like Osborne had prepared dozens of crystalline proteins and carefully studied their chemical composition and physical properties. They continued quietly to proceed on the assumption that proteins were definite large molecules.[88]

Svedberg, in the 1920s, at the University of Uppsala in Sweden and at the University of Wisconsin, set out to resolve this question. It was important to know how big proteins were and whether the size of the particles was uniform. In 1922, Svedberg and Rinde developed a method for determining the concentration distribution in a sedimenting column by light absorption.[89] While a visiting professor at the University of Wisconsin in 1923, Svedberg with Nichols constructed an optical centrifuge. In this instrument, the settling of the particles could be continuously observed or photographed during the run. However, the particles were carried down by sedimentation as well as by convection along the walls of the cell.[90] Upon Svedberg's return to Sweden in 1923, he and Rinde introduced sector-shaped cells to counter the problem of convection. It became evident that a new centrifuge had to be constructed, one providing a stronger centrifugal field than the optical centrifuge, in order to be used in the study of particles that could not be seen in light microscopes.

In July of 1924, the work with this new centrifuge had advanced so far that the first ultracentrifuge paper could be sent in for publication in the *Journal of the American Chemical Society.* Its title was "The Ultra-Centrifuge. A New Instrument for the Determination of Size and Distribution of Size of Particles in Amicroscopic Colloids." In the paper, the authors noted: "The new centrifuge constructed by us allows the determination of particles that cannot be made visible in the ultramicroscope. In analogy with the naming of the ultra-microscope and ultra-filtration apparatus we propose the name ultra-centrifuge for this apparatus."[71]

Increasing the speed of the ultracentrifuge and improvements of the design led to studies on protein, egg albumin and hemoglobin. The first calculated molecular weight of a protein from ultracentrifuge data was that of hemoglobin reported by Svedberg and Fåhraeus in 1926.[91] This value, 67,000, was also obtained by Adair in 1925 by osmotic pressure studies.[92] Adair's original value for hemoglobin was considered too high for a protein.

The value obtained by Svedberg and Fåhraeus was the result of a sedimentation equilibrium experiment. One of the conclusions reached was that for an analysis of the homogeneity or the polydispersity of the proteins under study, it would be necessary to use the sedimentation velocity method. In this procedure, the shape of the boundary in a dilute solution of a monodisperse substance would be determined solely by its diffusion coefficient. A larger spreading of the boundary than that corresponding to the diffusion of the substance, or the appearance of separate boundaries, would indicate inhomogeneity of the material.

Other improvements on the ultracentrifuge followed, especially in regard to the speed of the rotor. This led Svedberg to study other proteins, such as the chromoproteins, phycoertythrin, phycocyanin, and hemocyanin, and the serum proteins, albumin and globulin. As Professor Pedersen reported, "According to Svedberg the greatest sensation was the discovery of the giant hemocyanin molecules, isolated from the snail, *Helix pomatia*.[93] An estimate of the particle size showed that it had to be in the millions, and all the molecules had to have the same size. This was the first time such uniformly sized giant molecules had been observed."[94]

The question of the validity of proteins as large molecules suffered something of a setback when it became clear that the adjustment of pH could affect their size. To determine the charge property of proteins, Svedberg assigned his graduate student, Tiselius, the task of building an apparatus for the electrical separation of proteins. Such a construction, the Tiselius apparatus, became functional in 1937. In its main thrust, it complemented the search for homogeneity of proteins afforded by the ultracentrifuge. Later, it would serve for the separation of proteins and become indispensable for research on subunits of protein and nucleic acid. Svedberg's desire to put proteins in a distinct but measurable class of large molecules was apparently completed with the use of both the ultracentrifuge and the Tiselius apparatus. However, as noted by Williams, Svedberg went so far as to predict: "The X-ray (and possibly the electron-ray) analysis will ultimately give us a complete picture of the architecture of the protein molecule as it exists in crystals."[95]

The physical chemical characterization of proteins became a principal endeavor (Figure 2.11). Physical chemistry was founded in the later decades of the 19th century by such scientists as Ostwald, Van't Hoff, Arrhenius, Nernst, and others. It became recognized as a distinct discipline with the founding of the important journal, *Zeitschrift fur Physikalische Chemie*, in 1887 by Ostwald and Van't Hoff.[63]

2.3 Protein Structure

Generally, proteins are of two morphological classes, fiber-shaped and globular. Comprised of 20 amino acids linked together in a variety of sequences, proteins have a number of bonds and attractions which maintain the structure or spatial arrangements of their atoms. The spatial arrangement is referred to as a conformation. The Danish biochemist Karl Linderstrom-Lang, in 1924, proposed the terms primary, secondary, and tertiary to denote the structural hierarchy existing in proteins.[96]

The *primary structure* of proteins is the linear, covalently bonded amino acid sequence. In 1902 Emil Fischer and Franz Hofmeister, independently, advanced the hypothesis that in proteins the α-amino acid group of one amino acid and the α-carboxyl group of another amino acid are joined, with the elimination of a molecule of water, to form an amide linkage. An amide linkage joining two amino acids is termed a "peptide bond" or "peptide linkage." The product of this condensation reaction is termed a "peptide." Each amino acid unit of the peptide chain is referred to as an amino acid "residue."[97]

The process of peptide formation is repeated as the chain elongates. But the amino group of the first amino acid of a polypeptide chain and the carboxyl group of the last amino acid remain intact, and the chain is said to run from its amino terminus to its carboxy terminus.[96]

Insulin was the first protein in which the order of amino acid residues in the primary structure was determined. In the years 1945 to 1952, the English biochemist Sanger and collaborators determined the sequence of amino acids in this molecule. Insulin has a molecular weight of approximately 12,000. It consists of four polypeptide chains, of which two contain 21 amino acid residues apiece, and the other two contain 30.[98]

FIGURE 2.11 Scientists in attendance at the Pasadena Conference on the Structure of Proteins (1953). (Courtesy: The Archives, California Institute of Technology.)

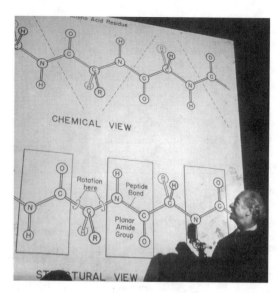

FIGURE 2.12   Linus Pauling (1901–1994) explaining his studies on the peptide bond. (Courtesy: The National Foundation March of Dimes.)

The *secondary structure* of proteins is concerned with the spatial arrangement of the polypeptide chain(s). Although fibrous proteins, generally, contain one type of secondary structure, globular proteins can incorporate several types of secondary structures in the same molecule.[72]

Two general types of conformation occur widely in proteins, the alpha helix and the beta sheet. In the late 1930s, Pauling and Corey undertook a study on proteins. Based on experimental data, especially the early work of the pioneer crystallographer, William Astbury (Figure 2.11), and exploiting the use of models, they analyzed the principal linkage in proteins, the peptide bond. They concluded that the C–N bond has a resonance nature — a partial sharing of two pairs of electrons between the carboxyl oxygen and the amide nitrogen. The amide C–N bonds are unable to rotate freely because of their partial double bond character — the oxygen atom has a partial negative charge and the nitrogen a partial positive charge, resulting in a small electric dipole. The C–N bond is one of the three covalent bonds intimately related to the peptide bond. The other two bonds are C_α–C and N–C_α.

Rotation is permitted about the N–C_α bond, the angle of rotation designated phi and also about the C_α–C bond, the angle of rotation designated psi (Figure 2.12). In principle phi and psi can have any value between –180° and +180°, but many values of phi and psi are prohibited by steric interference between atoms in the polypeptide backbone and amino acid side chains. Every possible secondary structure is described completely by the two bond angles phi and psi that are repeated at each residue[99] (Figure 2.12).

The Indian biophysicist Ramachandran first made calculations of sterically permissible regions of phi and psi.[100] Such plots are called Ramachandran plots. As shown in Figure 2.13, each point on a Ramachandran plot represents phi and psi values for an amino acid residue. Such data are retrieved from X-ray diagrams to high resolution. The amino acid glycine, with only a hydrogen atom as a side chain, can have a much wider range of conformation with respect to phi and psi than other residues.[100]

Stabilizing forces in proteins include covalent disulfide bonds and weak (noncovalent) interactions such as hydrogen bonds, and hydrophobic, ionic, and van der Waals interactions. They are defined as follows:[72] a *disulfide bond* is a covalent cross link between two polypeptide chains, formed by a cystine residue. *Hydrophobic interactions* are the association of nonpolar groups, or compounds, with each other in aqueous systems, driven by the tendency of surrounding water molecules to seek their most stable (disordered) state. A *hydrogen bond* is a weak electrostatic attraction between one electronegative atom, oxygen or nitrogen, and a hydrogen atom covalently linked to a second electronegative atom. *Weak ionic interactions* usually result from the effect of water on ionic compounds. Water dissolves salts such as NaCl

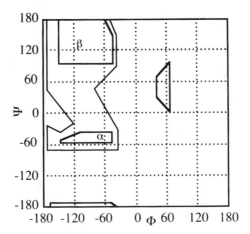

FIGURE 2.13 Ramachandran plot. Phi (Φ) and psi (ψ) refer to rotations of two rigid peptide units around the same C atom. The plot shows combinations of conformational angles phi and psi. Phi is the angle of rotation around the N–C_α bond and psi is the angle around the C_α–C bond from the same C atom. Areas labeled alpha and beta correspond approximately to conformational angles found for the usual right-handed helices and strands, respectively. (From Rhodes, G., *Crystallography Made Crystal Clear, A Guide for Macromolecular Models*, Academic Press, New York, 1993. With permission.)

by hydrating and stabilizing the Na^+ and Cl^- ions, weakening their electrostatic interactions and thus counteracting their tendency to associate in a crystalline lattice. The strength of these ionic interactions depends upon the magnitude of the charges, the distance between the charged groups, and the dielectric constant of the solvent through which the interactions occur. The dielectric constant is a physical property reflecting the number of dipoles in a solvent.

When two uncharged atoms are brought very close together, so that their surrounding electron clouds influence each other, random variations in the positions of the electrons around one nucleus may create a transient electric dipole, which induces a transient, opposite, electric dipole in the nearby atom. The two dipoles are weakly attracted to each other bringing the two nuclei closer. The force of this weak attraction is the *van der Waals* interaction. Although the four types of interaction are individually weak relative to covalent bonds, the cumulative effect of many such interactions in a protein or nucleic acid results in stabilization of the molecule.

The alpha helix was first described by Pauling, Corey, and Branson at the California Institute of Technology in 1950–1951.[101] The *alpha helix* is viewed as the simplest arrangement a polypeptide chain can assume. With its rigid C–N bond, but other bonds free to rotate, the polypeptide backbone is tightly wound along the long axis of the molecule, and the R groups (Figure 2.14) of the amino acid residues protrude outward from the backbone of the helix. The repeating unit is a single turn of the helix, extending about 0.56 nm along the long axis, matching closely to the periodicity Astbury observed on X-ray analysis of alpha keratin in hair. The bond angles Φ and ψ for the amino acid residues have values of $-60°$ and from $-45°$ to $-50°$, respectively. Each helical turn includes 3.6 residues. The twisting of the helix per turn in the α helix, corresponding to 5.4 Å or 1.5 Å per residue is commonly right handed. Stabilization of the alpha helix occurs through intra-hydrogen bonding, with each peptide bond along the chain participating in hydrogen bonding. The announcement of the alpha helix gained quick approval from X-ray diffraction patterns of hemoglobin crystals and keratin fibers, and was verified in the work of Kendrew in 1958 on high-resolution diffraction patterns of myoglobin.[96]

The second general type of conformation in secondary structure is the beta conformation or pleated sheet pattern. This conformation is a more extended one of the polypeptide chains, as seen in silk fibroin, a member of a class of fibrous proteins called beta-keratins. In the beta conformation, the backbone of the polypeptide chain is extended into a zigzag rather than helical structure. The hydrogen bonds can

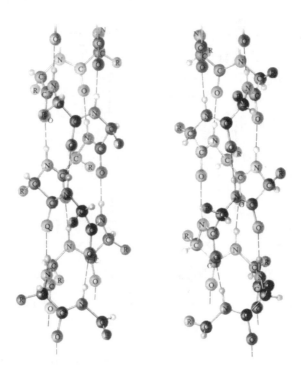

FIGURE 2.14 A drawing showing two possible forms of the alpha helix; the one on the left is a left-handed helix, and the one on the right is a right-handed helix. The right-handed helix of polypeptide chains is found in many proteins. The amino acid residues have the L configuration in each case. The circles labeled R represent the side chains of the various residues. (From Pauling, L., *General Chemistry*, Dover Publications, New York, 1988. With permission.)

be either intrachain, or interchain between the peptide linkages of adjacent polypeptide chains. The R groups of adjacent amino acids protrude in opposite directions from the zigzag structure, creating an alternating pattern.[72]

The beta pleated sheet can have either a parallel pattern, having the same amino to carboxyl polypeptide orientation, or antiparallel pattern, having the opposite amino to carboxyl orientation. The repeat period for the parallel structure is 0.65 nm and is 0.7 nm for the antiparallel arrangement (Figure 2.15). Beta strands can also combine into mixed beta sheets with some strand pairs parallel, and some antiparallel. Almost all beta sheets, parallel, antiparallel, and mixed, have their strands twisted in a right-handed sense. The secondary structures of α helices and β strands combine to build up structures. These secondary structures, connected by *loop regions* of various lengths and irregular shape, are at the surface of protein molecules. Loop regions that connect two adjacent antiparallel β strands are called *hairpin* loops.[94] Simple combinations of a few secondary structural elements having a specific geometric arrangement occur frequently in protein structures. These units are called *motifs* or super secondary structures. The simplest motif with a specific function consists of two α helices joined by a loop region.[96]

The three-dimensional arrangement of all atoms in a protein is referred to as the *tertiary structure*. Much of the stability required in a protein's tertiary structure is attained through weak hydrophobic interactions, involving nonpolar amino acid side chains in the tightly packed core of the protein. Tertiary structure is affected by amino acids that are farther apart in the polypeptide chain and these amino acids may interact when the protein is folded.[72]

The fundamental unit of tertiary structure is the domain. A domain is defined as a polypeptide chain or part of a polypeptide chain that can independently fold into a stable tertiary structure. The most frequent and most regular of the domain structures are the alpha/beta domains, which consist of a central parallel or mixed beta sheet surrounded by alpha helices. There are two main classes of alpha/beta proteins. In the first class there is a core of eight twisted parallel beta strands arranged close together,

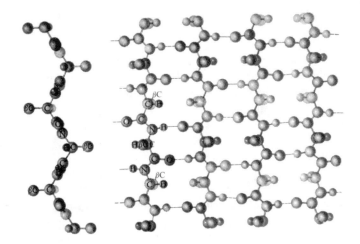

FIGURE 2.15 A drawing of the antiparallel-chain pleated sheet, a protein structure found for silk fibers. (From Pauling, L., *General Chemistry*, Dover Publications, New York, 1988. With permission.)

like staves, into a barrel. The alpha helices that connect the parallel beta strands are all on the outside of the barrel. The second class contains an open twisted beta sheet surrounded by alpha helices on both sides of the beta sheet. Both barrels and open sheets are built up from beta-alpha-beta motifs.[96]

Antiparallel beta structures comprise the second large group of protein domain structures. The beta sheets have a twist characteristic, and when two such twisted beta sheets are packed together, they may form a barrel-like structure or a saddle-shaped structure.[72] The polypeptide chain may also be wrapped around a barrel core, resembling a jelly roll. This motif has been found in the coat proteins of most of the spherical viruses examined by X-ray crystallography.[96]

Domains are also units of function. If the three-dimensional structure of a protein is altered the function of the protein may be affected. An extreme alteration of the three-dimensional structure results in denaturation, in which the loss of function usually results. Proteins are denatured by various agents, including heat, change in pH, organic solvents, and detergents. The process of denaturation results from a disruption of the weak interactions; however, not all of the stabilizing weak interactions need be disrupted for denaturation to occur.

Proteins denatured by various agents may regain their native structure and biological activity if they are returned to conditions in which the native conformation is stable. The reversal of denaturation is referred to as renaturation. Denaturation and renaturation are related to tertiary structure, which in turn is determined by the protein's amino acid sequence. Indications of reversible denaturation appeared in the 1920s, notably with the work of Hill on hemoglobin. He was able to separate the heme from the globin, reconstitute the two components, and demonstrate that the hemoglobin still had the original spectrum, and absorbed and released oxygen. Anson and Mirsky pointed out that in such experiments, globin was completely denatured when heme was separated and that the reconstituted protein did represent a regeneration of the native state and all of its properties. Similar denaturation-renaturation studies were noted: serum albumin by Mona, Spiegal, Adolph, Hsein Wu, Anson and Mirsky; typsin by Kunitz; pepsinogen by Herriott. All of this work was achieved by the 1920s and 1930s.[102]

In the 1950s, a clear-cut example of denaturation and renaturation was that of pancreatic ribonuclease, reported by Afinsen and White. Purified ribonuclease can be completely denatured by exposure in concentrated urea solution in the presence of a reducing agent, e.g., mercaptoethanol. The four disulfide bonds of ribonuclease are cleaved by the reducing agent to yield eight cysteine residues. The action of urea is to disrupt the stabilizing hydrophobic interactions, thus freeing the full polypeptide chain from its folded conformation. Under such conditions the protein loses its enzymatic activity and undergoes complete unfolding to a randomly coiled form. Removing urea and the reducing agent causes the randomly coiled, denatured ribonuclease to spontaneously refold into its correct tertiary structure, with full restoration of its enzymatic activity. This experiment proved that the amino acid sequence of the polypeptide chain of proteins contains all the information required to fold the chain into its native, three-dimensional structure.[102]

The result of the ribonuclease denaturation-renaturation study was followed by similar experiments with TMV protein. Preliminary studies pointed to the possibility of renaturation for the viral protein when Schramm and coworkers obtained subunits after treatment of the protein with alkali, and by Fraenkel-Conrat with acetic acid. Anderer demonstrated the complete renaturation of urea-denatured TMV protein, as did Fraenkel-Conrat, with guanidine-denatured TMV protein. These results further demonstrated that the amino acid sequence of the viral protein determined its conformation.[102]

The biological test for successful renaturation of TMV protein was its ability to aggregate at pH 4 to 6 to rods of identical diameter, and electron microscopic appearance as the intact virus. This was first observed by Takahashi and Ishii with excess TMV protein found in infected plants. The process was studied in greater detail by Schramm and Zillig in 1954, who tested the protein rod preparations for infectivity, finding them noninfectious.[102]

Quaternary structure is concerned with the interaction between subunits of proteins that are *oligomeric* — a multisubunit protein having two or more separate polypeptide chains or large protein assemblies. Hemoglobin is an example. The arrangement of proteins and protein subunits result in a three-dimensional complex. The interactions between subunits are stabilized by the same weak, multiple, noncovalent interactions that stabilize tertiary structure.[72]

2.4   Nucleic Acids

The discovery of nucleic acids was the result of investigations by Friedrich Miescher. In 1868–1869, Miescher, working in the laboratory of Hoppe-Seyler in Tubingen, isolated from the nuclei of pus cells a substance which he called "nuclein." This substance, a mixture, was acidic in nature and contained a relatively high phosphorus content. Up to this time the only known organic compound in tissue that contained phosphorus was lecithin. In time, Hoppe-Seyler's group isolated nuclein from other sources: egg yolk, yeast, casein, and the red cells of birds and reptiles. Collectively, these studies on "nuclein" were published in 1871 in Hoppe-Seyler's *Medicinisch-chemische Untersuchugen*.[103] Kossel's (Figure 2.16) research on "nuclein" showed that it contained a protein portion and a nonprotein portion, and nuclein became replaced with the term nucleoprotein. The nonprotein portion was termed nucleic acid by Altmann in 1889. When nucleic acids were broken down, Kossel found that among the products were purines and pyrimidines. He isolated two different purines, adenine and guanine, and a total of three different pyrimidines, thymine, cytosine, and uracil (Figure 2.17).[104]

The sugar in nucleic acid prepared from yeast was identified as D-ribose, a pentose sugar, by Levene in 1909, and confirmed by Gulland and colleagues in the 1940s (Figure 2.18). Levene and Mori identified the sugar in thymus DNA as a 2-deoxyribose. Thus, it was realized that there were two types of nucleic acid, DNA (originally called thymus nucleic acid) (Figure 2.19) and RNA (originally referred to as yeast nucleic acid).[104]

From Levene's work the formulas of the nucleotides, the small units out of which the large nucleic acid molecules were built up, were deduced. Todd, a Scottish chemist, and coworkers later synthesized all the naturally occurring nucleotide components of the nucleic acids, confirming Levene's studies.[104]

Biochemical studies on both RNA and DNA demonstrated that one of the acidic groups of a nucleotide phosphoric acid forms an ester with one of the hydroxyls of another nucleotide. Thus, in DNA the various nucleotides are joined and the internucleotide bond is the phosphodiester linkage. The DNA molecule has a long unbranched chain structure of the type that was originally postulated by Levene, and is shown in Figure 2.20. Since C-4′ in the deoxyribose sugar is occupied in ring formation and C-2′ carries no hydroxyl group, only the hydroxyl groups at positions 3′ and 5′ in the sugar residue are available for internucleotide linkages. Treatment with appropriate enzymes break down this structure to nucleoside 3′-phosphates or to nucleoside 5′-phosphates. RNA is a polymer, the monomer units of which are

FIGURE 2.16 Albrecht Kossell (1853–1927). (Courtesy: Department of Special Collections, Van Pelt-Dietrich Library Center, The University of Pennsylvania.)

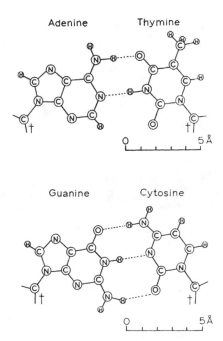

FIGURE 2.17 Complementary base pairing in DNA of the purine adenine and the pyrimidine thymine, and of the purine guanine and the pyrimidine cytosine.

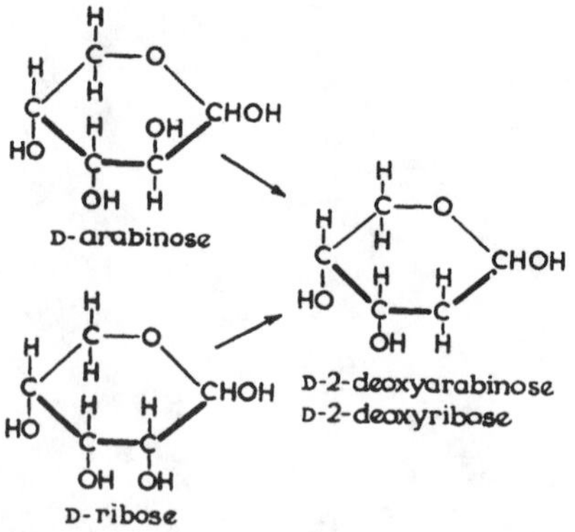

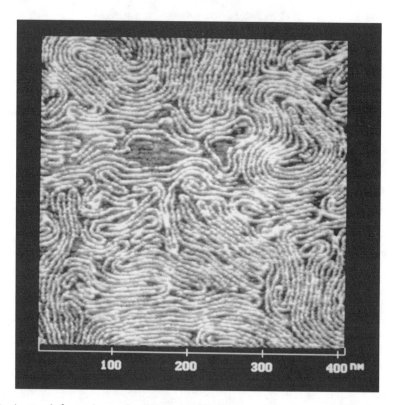

FIGURE 2.18   Pentose sugars showing the structure of D-ribose and D-deoxyribose.

FIGURE 2.19   An atomic force microscopic 409 nm scan of calf thymus DNA adsorbed as a close-packed monolayer to a mica-supported cationic lipid bilayer and imaged in buffer solution using the Tapping Mode. (With the permission of Digital Instruments, Santa Barbara, CA.)

ribonucleoside monophosphates. The main internucleotide linkages are phosphodiester groups connecting C-5′ in one nucleotide with C-3′ in the next nucleotide.[104]

Indications of the giant size of the nucleic acids was realized by Miescher, who noted the slow diffusivity of solutions of these substances as well as their great lability. With the emphasis on purity, investigators used harsh extraction procedures and thereby obtained only portions of nucleic acids of relatively small size. However, Olby notes that from 1935 to 1940, reports of molecular weights for DNA changed from being considered low to that of giant molecules. DNA and RNA became recognized as large molecular

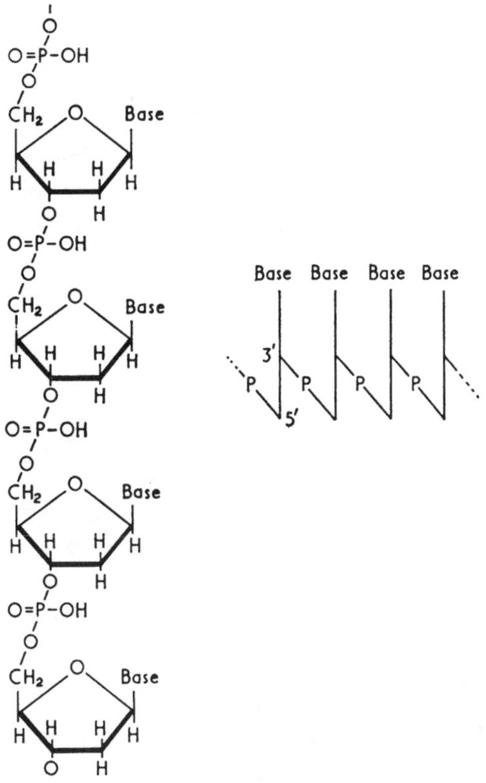

FIGURE 2.20 A portion of DNA showing internucleotide linkages.

species.[48] Generally, viral DNA varies in molecular weight from a little over 10^6 to more than 10^8. The molecular weight of viral RNA varies from about 10^6 to more than 10^7.

As noted above, of the two viral components, protein, for the first three to four decades of the 20th century, was considered the primary substance conveying hereditary information. Yet, from the 1940s, studies pointed to nucleic acid as the genetic molecule.

The first direct evidence came in the 1940s through a discovery made by Ostwald T. Avery, Colin MacLeod, and Macyln McCarty.[23] These investigators found that DNA extracted from a virulent strain of the bacterium *Streptococcus pneumoniae* genetically transformed a nonvirulent strain of this organism into a virulent form. It was concluded that the DNA extracted from the virulent strain carried the inheritable genetic message for virulence. Their proof rested on observations that treatment of DNA with proteolytic enzymes did not destroy the transforming activity, but treatment with deoxyribonuclease, a DNA-hydrolyzing enzyme, did.

The second experiment establishing DNA as the genetic molecule was reported in 1952 by Alfred D. Hershey and Martha Chase.[19] These investigators used radioisotopes to follow the path of T2 phage infecting its bacterial host (*Escherichia coli*). Radioactive phosphorus (^{32}P) was used to label the nucleic acid of the virus and radioactive sulfur (^{35}S), the protein component. The results demonstrated that when the T2 virus infects *E. coli*, it is the phosphorus-containing DNA of the viral particle, not the sulfur-containing protein of the viral coat, that enters the host cell, commandeering it to manufacture viral particles.

Hershey and Chase were able to follow the viral infection of the host bacterium with the electron microscope. Phage attached to the bacterium signaled that infection had begun. Using a waring blender, attached phages could be sheared away and removed through centrifugation. With the viral progeny radioactive with ^{32}P, indicating that nucleic acid had been injected into the host, the conclusion could be reached that the protein of the phage was on the outside of the viral particle and the nucleic acid within.

A third confirming experiment of the infectivity of nucleic acid in microbes was that of proving that the isolated nucleic acid was the infectious entity. In the 1950s, Gierer and Schramm at Tubingen and Fraenkel-Conrat and colleagues at Berkeley independently carried out such studies on TMV nucleic acid. The separated viral RNA was infectious, and elicited the typical disease symptoms characteristic for the virus strain. The protein of the progeny of such RNA infection was identical in composition to that of the strain of origin, even when the RNA had been reconstituted with proteins from another stain. Thus, the virus coat protein plays only a protective and infectivity-enhancing role, but the genetic function of the virus is carried entirely by its RNA. When the virus protein was allowed to reassemble in the presence of viral RNA at neutral pH, RNA-containing particles, indistinguishable from the 300 nm rods, were observed. Such reconstituted virus was also as stable over a wide pH and temperature range as the original virus, in contrast to the rods of protein alone, which could be of any length, but were only stable near the isoelectric point of the protein. The RNA was able to stabilize and thereby determine the length of the virus particles.[102]

2.5   Nucleic Acid Structure

A number of analytical procedures led to the establishment of the structure of DNA. The most powerful of these procedures is the use of X-ray crystallography. The first X-ray photographs of fibrous DNA by Astbury and Bell[105] displayed a very strong meridional reflection at 3.4 Å distance, implying that the incorporated bases were stacked upon one another. Some conclusions from this work were basic to an understanding of the structure of DNA: (1) the nucleotides do not follow each other always in the same order, fostering great variation; (2) the nucleotides form a single-chain molecule, with the sugar and base rings coplanar and at right angles to the axis of the chain. At a Cold Spring Harbor Symposium on the chemistry of proteins, at which some of this work was discussed, Stuart Mudd commented:[106] "So it seems apparent that the pile of nucleotides by slight changes in the order in which nucleotides occur, or possibly by other changes in configuration, give us an adequate basis for specificity." Another highly important study involving X-ray crystallography was the study by Furberg in 1949 on the crystal structure of the nucleoside cytidine. This work resulted in the first description of the molecular structure of a nucleoside.[107]

Early diffraction photographs of DNA fibers taken by Franklin (Figure 2.21) and Gosling[108] and by Wilkins and colleagues[109] in England, revealed two types of regular DNA structures: A-DNA and B-DNA. The A-DNA form is obtained under dehydrated, nonphysiological conditions. The B-DNA form is obtained when DNA is fully hydrated, as it is *in vivo*. The B form of DNA revealed a helical structure by X-ray diffraction. Later, improvements in the methods for the chemical synthesis of DNA have made it possible to study single crystals of short DNA molecules of almost any selected sequence. Such improvements have essentially confirmed the diffraction models of the A- and B-DNA forms. Moreover, from such studies a new structural form of DNA, called Z-DNA, has been discovered.[110]

Chemical studies would also provide important clues to nucleic acid structure. From Todd's laboratory it was established that nucleotides are linked via 3,5-phosphodiester bonds to produce a linear polymer of DNA. From electrotitrimetric studies, it was concluded that the bases of DNA were linked by hydrogen bonding.[103]

The development of paper chromatography in 1944 by Martin and Synge was utilized by Erwin Chargaff and coworkers to determine the quantity of each of the nitrogenous bases present in a particular nucleic acid molecule. Tests on DNAs from a variety of sources showed that in general the number of adenine units in each was equivalent to the number of thymine units, while the number of guanine units was equivalent to the number of cytosine units.[111]

From these and other basic studies, Watson and Crick (Figure 2.22) elucidated the structure of B-DNA.[112] They depicted the structure as composed of hydrogen-bonded base pairs of adenine with thymine (A-T) and guanine with cytosine (G-C) to explain the chemical data. Physically these base pairs are stacked like a role of coins at 3.4 Å distance, as shown by Astbury and Bell. A right-handed rotation about 36° between adjacent base pairs produces a double helix with 10 base pairs per turn. The average helical twist angle is 33.9° and the spacing along the helical axis from one base point to the next is 3.4 Å.

FIGURE 2.21 Rosalind Franklin (1920–1958). (Courtesy: Cold Spring Harbor Laboratory Archives.)

FIGURE 2.22 Frances Crick and James D. Watson (1953). (Courtesy Cold Spring Harbor Laboratory Archives.)

Utilizing Furberg's data on the structure of cytidine, dimensions and shapes of individual nucleotides were constructed, resulting in a model of the helix with bases located along the helix axis (inside) and sugar-phosphate backbones winding in antiparallel orientation along the periphery. Later work with DNA polymerase provided experimental evidence that the strands are indeed antiparallel. This fact was also confirmed by X-ray crystallography.

A-DNA is also a right-handed helix. The rails (strands) are the two antiparallel phosphate-sugar chains and the rungs are purine-pyrimididine base pairs, which are hydrogen bonded to each other. In A-DNA there are an average of 10.9 base pairs per turn of the helix, corresponding to an average helical twist angle of 33.1° from one base pair to the next. The spacing along the helical axis from one base pair to the next is 2.9 Å.[72]

The spatial relationship between the DNA strands creates a wide, major groove and a narrow, minor groove between the strands. In B-DNA the helical axis runs through the center of each base pair. The base pairs are stacked nearly perpendicular to the helical axis, and the major and minor grooves are of similar depths. In A-DNA, the helical axis is shifted from the center of the bases into the major groove, bypassing the bases, and the base pairs are not perpendicular to the axis but are tilted between 13° and 19°. This arrangement makes the major groove very deep, extending from the surface all the way past the central axis and part of the way out toward the opposite side, while the minor groove is shallow, scarcely more than a helical depression. The edges of the base pairs in DNA that are in the major groove are wider than those in the minor groove, arising from the asymmetrical attachment of the base pairs to the sugar-phosphate backbone. These edges contain different hydrogen donors and acceptors for potential specific interaction with proteins.[96]

In Z-DNA, the helix is left-handed and the sugar-phosphate backbone follows a zigzag path. It has a thin and elongated helical shape. Z-DNA has a deep but very narrow minor groove and the major groove seen in other DNA forms has been pushed to the surface so that it is no longer a groove at all.[96]

It was originally believed that hydrogen bonds between the base pairs in the DNA double helix were mainly responsible for the stability of the helix both in the solid state and in aqueous solution. It is now agreed, however, that the contribution made by hydrogen bonding is relatively slight and that the major factors responsible for helix stability are what are known as base-stacking forces, the hydrophobic interactions between the heterocyclic bases as they stack vertically in ordered helical array.[104]

RNA usually occurs as a single-stranded molecule with a right-handed helical conformation. Base-stacking interactions dominate the structure of RNA. The bases in RNA are adenine, uracil, cytosine, and guanine. RNA can base pair in a complementary fashion with strands of RNA or DNA. The standard base-pairing rules apply: adenine with uracil (RNA strand) or thymine (DNA strand), and cytidine with guanine. What is sometimes observed is base pairing between guanine and uracil, when two strands of RNA become joined. The paired strands in RNA or between RNA and DNA are antiparallel, as in DNA. Weak interactions, especially base-stacking hydrophobic interactions, play a major role in stabilizing RNA structure. When complementary union with other nucleic acid strands occur, RNA or DNA, the predominant double-stranded structure is an A-form, right-handed double helix. The B-form of RNA has not been observed. Z-form helices involving RNA have been prepared in the laboratory under conditions involving the presence of high salt concentration or high temperature conditions.[72]

2.5.1   Denaturation of Double-Stranded Nucleic Acids

Isolated, native DNA is highly viscous at pH 7.0 and at 20 to 25°C. However, at extremes of pH or at temperatures above 80 to 90°C, a noticeable drop in viscosity occurs, indicating that DNA has undergone a physical change. This physical change is a result of the denaturation of DNA, as expressed in an unwinding of the two strands. The unwinding of the strands is a consequence of the breakdown of the stability factors holding the strands together: disruption of the hydrophobic interactions between the stacked bases and a breaking of the hydrogen bonds. It should be noted that the denaturation of DNA at this level involves no breaking of covalent bonds.

Strand-separated DNA can be renatured so that its separated strands connect. If the temperature or pH is returned to normal values, a spontaneous winding in helical fashion of the two strands takes place. Each species of DNA has a characteristic denaturation temperature, referred to as the melting point. For DNA rich in G-C base pairs, a higher melting point is observed than for DNA rich in A-T pairs. The G-C base pairing, with three hydrogen bonds, gives its DNA more stability than do the two hydrogen bonds

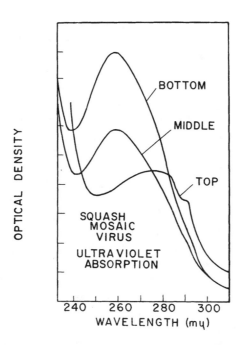

FIGURE 2.23 The nucleoprotein nature of viruses as demonstrated by absorption curves of the bottom, middle, and top components of squash mosaic virus. Curves of the bottom and middle components, nucleoproteins, show an absorption at about 260 μm. The top component, which is noninfectious, contains protein but no nucleic acid. It has an absorption maximum at about 280 μm, as is common for proteins. (Data from Reference 257.)

present in A-T pairs. Therefore, more heat energy is required to dissociate C-G-rich DNA than is needed for A-T DNA.

There is a relationship between the melting point of a species DNA and its C-G ratio, provided that pH and ionic strength conditions are fixed in the determination of the melting point. Viral and bacterial DNA base ratios, as well as other species of DNA, have been analyzed by the melting point determination.

Denaturation by heating or pH changes also occurs with double-stranded RNA and with duplexes composed of one strand of DNA and one strand of RNA. Double-stranded RNAs are more stable than DNA duplexes. A temperature higher than that at which DNA will denature is necessary to denature duplex RNA. A duplex of DNA and RNA is generally intermediate between that of double-stranded RNA and DNA. Why such differences occur has not been determined.[72]

2.6 Viruses as Nucleoproteins

Following the report by Bawden and Pirie in 1937 that TMV contained nucleic acid in addition to protein, Max Schlesinger, in the same year, reported that purified preparations of bacteriophage consisted mainly of nucleic acid and proteins. It became reasonable to assume that viruses were nucleoproteins (Figure 2.23). In this connection, Astbury proclaimed: "To the molecular biologist there can be no question but that the most thrilling discovery of the century is that of the nature of the tobacco mosaic virus: it is but a nucleoprotein."[113]

The ultracentrifuge was applied to the study of viruses. In 1936 Eriksson-Quensel and Svedberg obtained sedimentation coefficient values for TMV ranging from 190 to 235 S.[114] A mean molecular weight for the virus of 17 million was calculated, based on sedimentation equilibrium measurements. While this value is lower than the now accepted value of approximately 40 million, the study spurred interest in the use of the ultracentrifuge in virus studies.

Determining the molecular weight of the viruses with the ultracentrifuge was initially difficult. The sedimentation coefficient had to be related to the diffusion coefficient, which was not an easy parameter to measure for large molecules. In 1938, Lauffer used viscosity data to obtain the length:breadth ratio of TMV, which combined with an S value of 174, obtained a reliable molecular weight value of 42.5 million.[115]

When viruses were first isolated it was assumed that the active particles were spherical. However, Takahashi and Rawlins in the early 1930s derived a rod shape for TMV from birefringence studies.[116] These observations were confirmed from the work of Lauffer and Stanley.[117] The shape and size of viruses would be obtained readily with the introduction of a new instrument, the electron microscope.

Electron optics is considered to have originated in 1926, when Busch first demonstrated the existence and properties of electron lenses. The electron microscope, invented by Knoll and Ruska in 1932,[9] provided a direct means of measuring the size of viruses as well as of obtaining their shape. Based on the wave nature of the electron postulated by de Broglie in 1925, and proved experimentally by Davisson and Germer in 1927, the limit of resolution was theoretically determined as about 0.22 nm (2.2 Å).

In 1939, the electron microscope revealed the long, thin particles of tobacco mosaic virus. Kausche et al., in 1939,[118] and Ruska, in 1941,[119] reported that phages were sperm-shaped. In the U.S., the RCA Company, headquartered in Camden, New Jersey, was quick to develop this instrument. Thomas F. Anderson was one of the first to investigate viruses with the electron microscope.[120] Having obtained a National Research Council Fellowship with the RCA Corporation in Camden, New Jersey, in 1940, Anderson came in contact with electron microscope pioneers: Zorykin, in whose laboratory the electron microscope was housed, and Marton, who had built an electron microscope capable of 50 Å resolution and had taken the first pictures of a biological specimen, that of bacteria. Anderson engaged in collaborative studies with Stanley on tobacco mosaic virus and with Delbruck and Luria on bacteriophages.[121]

With the establishment of the Phage School at Cold Spring Harbor in the 1940s, Delbruck, Luria, and colleagues set out to study seven phages infecting the *Escherichia coli* bacterium: T1–T7 (T for type). This set of phages had been collected by Demerec and Fano for studies of the patterns of mutation of *E. coli*, strain B, to resistance to the phages. The T phages give easily countable plaques.

Some basic questions were asked, ones that would serve as a basis of molecular biology: "How could the virus multiply inside the bacterium without ever entering its host?. "Where in the virus is the nucleic acid?"[48]

From electron microscope studies the seven T phages could be classified into four morphological groups: (1) T1; (2) T2, T4, T6; (3) T3, T7; and (4) T5. An important point concerning the attachment by phages to host bacteria — by heads or by tails — was cleared up with critical point drying and electron microscopy. The electronmicrographs repeatedly showed phages attached by their tails to the host bacteria.[59] Certain conclusions were reached: (1) characteristic particles are always present in highly active phage suspensions and missing in any control suspensions (media, bacterial cultures, bacterial filtrates, etc.); (2) they are readily absorbed by the bacterial cells of the susceptible strain and fail to be absorbed by other bacteria; (3) the size from a given strain is uniform and corresponds essentially to measurements by indirect methods; (4) the structure of both the "head" and "tail" is characteristic of the strain of phage; (5) preliminary experiments on the lysis process seem to demonstrate the liberation of these particles from the lysing bacteria.[121]

The infectious process involving T phages and host bacterium was followed utilizing the electron microscope. The latent period occurring in phages could be timed, and the burst of new phages out of the host could be observed. As noted, the question of the infectious component of the virus, nucleic acid or protein, or both was resolved with the classic work of Hershey and Chase with radioisotopes, ^{35}S and [32]P. The electron microscope played an important role in such experiments.

Another experiment demonstrating that DNA is contained within the phage was conducted by Anderson[122] and by Herriott.[123] When T4 phage was incubated in concentrations higher than 2 *M* and then rapidly diluted in a medium of low osmotic pressure, the electron microscope showed virus particles with empty heads. This work demonstrated that such empty-headed phage particles consisted of protein

FIGURE 2.24   D.L.D Caspar (right) with J. Wyman (1971). (Courtesy: Cold Spring Harbor Laboratory Archives.)

and contained none of the DNA present in normal T4. In this way progress was made in elucidating the functional makeup of the T-even bacteriophages.

Improvements in electron microscopic techniques continued and greatly aided the investigation of viruses. Mention was made of critical point drying, by which water is gently removed from the sample application. The technique of shadow casting, in which heavy metals are sprayed under vacuum onto the object to be viewed, gave a clearer picture of virus particles as they cast a shadow of their shape.[124] This procedure was enhanced with the technique of negative staining whereby a drop of a heavy metal solution could be applied to the sample on an electron microscope grid, resulting in a much improved contrast of virus and background.[125]

2.7   Physical Principles in the Construction of Regular Viruses

Regular viruses consist of a nucleic acid component of high molecular weight, which is responsible for infectivity, and a protective package of protein, also of high molecular weight, enclosing the nucleic acid. The nucleic acid, DNA or RNA, directs the synthetic machinery of the host to produce more virus particles. A complex DNA virus or a retrovirus may involve the synthesis of infection requiring protein enzymes, as well as the synthesis of its viral proteins.

As noted, the nucleic acid of viruses has a large range of molecular weight. The DNA of Tipula iridescent virus and of vaccinia virus is about 150×10^6. This value is many times greater than the DNA of the small pleura-pneumonia-like organism, PPLO, now classified as a bacterium, or of the small tobacco necrosis satellite virus particles. The satellite particles, with relatively small nucleic acid size, do not carry enough information for the multiplication of the particles. Such particles can only reproduce in association with the larger tobacco necrosis virus, whose DNA molecular weight is much higher.[126]

Caspar and Klug (Figures 2.24 and 2.25) proposed the construction of viruses as a self-assembly process in which the identical subunits of protein may or may not be exactly identical environments.[127] These two possibilities result in *equivalent* or *quasi-equivalent* architecture, respectively, and is demonstrated in the two common forms of viruses, helical rods or rod-like structures and the icosahedral viruses. In the case of the helical rod structure, TMV is the best example of self-assembly,[128] and the process may be observed *in vitro*.[129]

FIGURE 2.25   Aaron Klug, second from right, with clockwise: R.G. Shulman, J. Hopfield, R.T. Ogata, and P. J. G. Butler (1971). (Courtesy: Cold Spring Harbor Laboratory Archives.)

The protein subunits of TMV are closely packed in a helical array, whereby any one subunit is structurally, as well as chemically, indistinguishable from any other. The RNA chain is coiled in a compact way between the turns of the protein helix, and the phosphate-sugar backbone is folded so that the nucleotides are equivalently related to each protein subunit.

To reacquaint the reader with terms, morphological units composing the shell have been given the name *capsomeres*. The shell itself is the *capsid*. The region inside the capsid is the *core*. The outer membrane, seen surrounding the capsid of some viruses, is the *envelope*.

In the case of TMV, the protein subunits have the capacity to assemble themselves, with or without nucleic acid, to form the framework of the packaging. This property of self-assembly is intimately connected with the helical symmetry of the virus. The result of the structure of the protein component is a packing of identical units so that the same kinds of contact are available continuously. In the final structure, each subunit, except for those at the ends, makes the same bonds — each subunit then is in the same environment. The subunits are *equivalent,* and the resulting structure is symmetrical. The length of a helical virus, such as TMV, is not determined by its symmetry nor by its geometrical parameters. As noted, the length of the helical array of protein in the intact virus particle is determined by the length of the RNA chain.

Flexible or filamentous rod-shaped viruses are helical but are not held rigid by strong interactions between successive turns of the helix. The subunits in a sinuous helical virus are not exactly equivalent, as this restriction would demand a straight axis of symmetry. However, since the local bonding pattern would not be changed very much in the bending of the axis, the subunits in a flexible helical structure can remain *quasi-equivalently* related, e.g., in icosahedral viruses.

From X-ray diffraction studies and electron microscope observations, a number of facts were realized in terms of structure. Crick and Watson[130] pointed out that with an isometric particle, only cubic symmetry would be possible. Three types of cubic symmetry exist: tetrahedral, which has 12 identical subunits; octahedral, which has 24 identical subunits; and icosahedral which has 60 identical subunits. In virus structure, the icosahedral symmetry appears to be preferred. The first experimental evidence for icosahedral symmetry in a virus came from the X-ray diffraction studies on tomato bushy stunt virus (TBSV)[131] and turnip yellow mosaic virus.[132] These investigations confirmed the prediction of Crick and Watson that some viruses possess cubic symmetry and are icosahedrons. Icosahedral symmetry was also shown for Tipula iridescent virus[133] and polio virus.[134] Thus, it seemed fairly certain that icosahedral symmetry exists in quite unrelated viruses.

Therefore, a general principle appears. Icosahedral symmetry allows for the use of the greatest possible number of identical symmetric units, namely 60, to build a spiral framework in which they are also identically packed. Icosahedral symmetry would also appear to be the most efficient form of packing. X-ray diffraction studies on turnip yellow mosaic virus[135] and tomato bushy stunt virus[131] led to a conclusion by Caspar and Klug concerning icosahedral viruses. If the protein shell has strict icosahedral symmetry, then it must be built up of either 60 subunits, or a multiple of 60.

The smallest icosahedral viruses have a shell of outer diameter of 150 to 200 Å, and would take up to 60 units of 20,000 molecular weight. The next largest diameter commonly found in the small viruses is about 280 Å. To occupy this shell size, from about 150 to 250 of the sized subunits would be required, so that a multiple of 60 is required for icosahedral symmetry. Yet, turnip yellow mosaic virus and wild cucumber mosaic viruses are icosahedral but their protein shell is built up of more than 60 identical subunits. It is impossible, in line with Caspar's and Klug's theory, to put more than 60 identical units on the surface of a sphere in such a way that each is identically situated. If more than 60 units are put on such a surface, the members cannot be equivalently related. Caspar and Klug suggested that while the shell is held together by the same type of bonds throughout, these bonds may be deformed in slightly different ways in the different, non-symmetry-related environments. In the case of helical rods, such exceptions exist. The dahlemense strain of TMV departs slightly from equivalent packing of identical units, so, for the icosahedral viruses, a departure from the requirement of 60 subunits necessary to fill the shell would involve a *leeway* in arrangement, allowing for a degree of nonequivalence in subunit packing.[136]

The simplest virus known, the satellite virus of tobacco necrosis virus, has 60 identical subunits in its shell. However, as noted, the satellite is not a self-sufficient virus — it does not encode all the functions required for its replication. Such particles require a helper virus, the normal tobacco necrosis virus, to supply functions it does not encode.

Self-sufficient viruses have more than 60 subunits in their shell. Caspar and Klug maintained that for viruses requiring more than 60 shell subunits, only certain multiples could be accommodated while still preserving the icosahedral shape. These multiples, called triangulation numbers, T, are 1, 3, 4, 7 … of 60 subunits, are those likely to occur ($T = h^2 + kh + k^2$, where h and k are any integers). When T is larger than 1, it is no longer possible to pack the protein subunits into the icosahedral shell in a strictly equivalent way as it is for a $T = 1$ structure like satellite necrosis virus. In this virus, all subunits then have the same environment and the same packing interactions.

If there are greater than 60 subunits, identical or different, it is possible to pack the subunits with only slightly different bonding patterns, in a quasi-equivalent way. Again, the multiples of 60 have certain specific values of 1, 3, 4, 7 ….

In 1978, Harrison and coworkers[137] first demonstrated for a spherical virus that the tomato bushy stunt virus exhibited T3 quasi-equivalent arrangement of its subunits. The structure of TBSV was determined to 2.9 Å resolution with X-ray scattering, and is made up of 180 identical subunits.

What is the packing for spherical viruses that do not have identical subunits? The picornaviruses have a protein shell that contains four different polypeptide chains and the RNA is single stranded. These viruses have a molecular mass of approximately 8.5 million daltons. The RNA is a long molecule with a molecular mass of about 2.5 million daltons occupying the interior of the spherical virion. The arrangement of the subunits in the shell of picornaviruses is similar to that in $T = 3$ plant viruses. The three-dimensional structure of various members of the Picorna group of viruses has been determined. Rossman and coworkers at Purdue University[138] determined the structure of Mengo virus to 3.0 Å resolution and that of human rhino virus, strain 14, to 2.6 Å resolution; James Hogle and his group at Scripps Clinic, La Jolla, California determined the structure of polio virus to 2.9 Å resolution;[139] and Stuart and coworkers at Oxford University determined the structure of foot-and-mouth disease virus to 2.9 Å resolution.[140]

All the above viruses have the same topological subunit type of structure, namely a jelly roll topology, noted above. However, in 1986, Valegard, in the laboratory of Liljas[141] at Uppsala, reported that the

FIGURE 2.26   Buckminster Fuller (1895–1983) with a model of his geodesic dome. (Courtesy: Buckminster Fuller Institute, Santa Barbara, CA.)

bacteriophage MS2 with a T3 packing of subunits, had a different fold of the subunit. The phage, determined to 3 Å resolution, belongs to a family of small single-stranded RNA phages, MS2, R17, f2, and Qβ, that infect *E. coli.* These phages are about 250 Å in diameter. The subunit of MS2 phage folds into a five-stranded up-and-down antiparallel β sheet with an additional short hairpin structure at the amino terminus and two α helices at the other end of the chain. The α helices are responsible for interactions with a second subunit to form a tight dimer. The two β sheets of each subunit in the dimer are aligned at their edges so that a continuous β sheet of 10 adjacent antiparallel β strands is formed with α helices from one subunit packing against the β sheets of the second subunit. This subunit structure is different from the jelly roll structure found in all other spherical viruses so far.[96]

A nucleic acid cannot code for a single protein molecule that is large enough to enclose it. The protein structures in virus coats (capsids) generally function simply as enclosures. The simplest viruses have just one type of polypeptide chain, which forms either a rod-shaped or a roughly spherical shell around the nucleic acid. The rod-shaped tobacco mosaic virus is a right-handed helical filament with 2130 copies of a single protein that interact to form a cylinder enclosing the viral RNA. Other examples of simple viruses are the spherical satellite tobacco necrosis virus, tomato bushy stunt virus, and southern bean mosaic virus.

The underlying theory of how viruses are structured, as seen by X-ray diffraction studies, owes much to Buckminster Fuller's concept of geodesic domes (Figure 2.26). In this connection, Caspar and Klug, in their article on Physical Principles in the Construction of Regular Viruses,[127] acknowledge their stimulating discussions with Fuller and the availability of his unpublished notes.

3

Purification
Considerations

In the study of viruses, pure preparations are needed, and in some cases relatively high concentrations of the virus must be obtained. Generally, each virus preparation has its own set of preliminary steps leading to purification and concentration, for example low speed centrifugation, pH adjustment, and salting-out techniques. Such procedures will not be discussed here. Rather, this section is concerned with the enrichment of the virus preparation just prior to chemical and physical analyses.

A number of procedures are employed to increase the quantity of pure virus, including centrifugation, electrophoresis, and chromatography. However, preparative centrifugation techniques are extensively used and will be the main concern of this discussion.

Some of the centrifugation techniques described below permit the determination of buoyant density. In addition, sedimentation coefficients have been estimated and approximate molecular weights of the virus calculated. These parameters are discussed in detail in Chapter 5.

3.1 Differential Centrifugation

This procedure involves spinning a sample at alternating low speeds, 10,000 to 20,000 rpm, and high speeds, 30,000 to 50,000 rpm. A rotor speed is usually chosen to pellet the contaminating large components first and the virus second, leaving in the supernatant, material that is smaller sized than the virus particles (Figure 3.1). Repeated cycles of low-speed and high-speed centrifugation improve the purity of the virus, but reduce the yield with each cycle (Figure 3.2). Differential centrifugation was developed as an effective purification procedure for tobacco ringspot virus and tobacco mosaic virus by Stanley and Wyckoff in 1937.[142] The reason that this method, with its apparent low resolution, has worked so well is that it has been most often applied to the separation of particles whose sedimentation coefficients differ by orders of magnitude. It has been least successful in cases where very small differences in sedimentation rate exist.

3.1.1 Zonal Centrifugation

To achieve higher resolution than that afforded by differential centrifugation, two general types of centrifugation have been developed. In the first type, particles are separated into discrete zones on the basis of differences in sedimentation rate. This technique is called rate zonal centrifugation. In the second type, separation of particles is based on differences in buoyant or banding density. This technique is called isopycnic-zonal centrifugation. In each procedure a gradient is employed to counter convection.

3.1.1.1 Rate Zonal Centrifugation

A swinging bucket rotor (Figure 3.3) is generally used in this technique, originated by Brakke[143] in the early 1950s. A centrifuge tube is prepared, containing a continuous density gradient, usually sucrose, formed either by diffusion between discontinuous layers or by a gradient mixing device. At rest, a thin zone of the viral suspension to be fractionated is layered on top of the gradient.

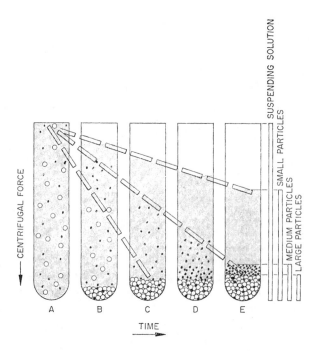

FIGURE 3.1 Sedimentation of a particular suspension in a centrifugal field. Initially all particles are uniformly distributed through the centrifuge tube (A). As centrifugation proceeds (B) all particles present sediment at their respective sedimentation rates, resulting first in the total sedimentation of the species with the largest particles (C). The degree of cross contamination in the pellet between two species of particles at the moment when one species has sedimented completely is approximately proportional to the ratio of the sedimentation rates. Further centrifugation (D) results in complete sedimentation of intermediate-sized particles (E). The distribution of particles in tube E is shown in the bar graph at the right. (Courtesy: National Cancer Institute.)

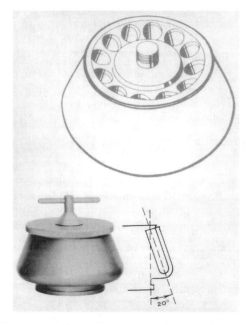

FIGURE 3.2 An angle-head rotor commonly used in preparative centrifugation. (Courtesy: Beckman Instruments.)

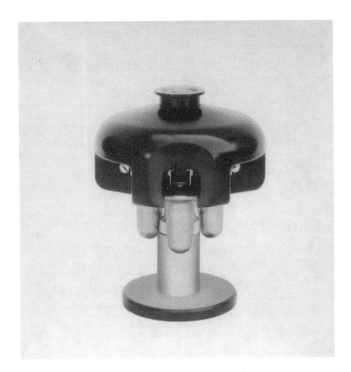

FIGURE 3.3   Swinging bucket rotor. (Courtesy: Beckman Instruments.)

The centrifuge tube is accelerated slowly so that the transition from rest to a horizontal position of the tube occurs without disturbing the gradient or the sample layer. Under the action of the centrifugal field, the viral particles will migrate along the axis of the tube. Particles are separated according to size and shape. After the required separation has been made, the tubes are decelerated to rest and scanned for zones separated from one another. Recovery of zones may be achieved by puncturing a hole in the bottom of the tube and collecting the contents (Figure 3.4).

The amount of material that may be separated with conventional high-speed swinging bucket rotors is very small. For the SW 39 rotor the gradient is 4 ml per tube, and the rotor holds three tubes. The sample layer in this case is only a fraction of a milliliter. However, sufficient material may be collected for chemical analysis or for further fractionation studies.

3.1.1.2   Isopycnic Centrifugation

Isopycnic centrifugation involves spinning particles in a gradient solution until they reach their isodensity level, i.e., where the gradient density and the buoyant density of the particles are equal. At this isodensity condition, no further sedimentation will occur, and the particles may be considered to be at equilibrium (Figure 3.5).

Isopycnic-zonal centrifugation was originally used in the 1930s to stratify, *in situ,* the subcellular particles of sea urchin eggs and to separate the eggs into light and heavy halves.[144]

In 1957, Meselson, Stahl, and Vinograd employed the technique in their classical studies on zonal separations of DNA in cesium chloride gradients.[145]

Angle head rotors and large swinging bucket rotors may be used for purifying large quantities of virus by isopycnic techniques. The most commonly used gradient-forming solutes are sucrose and salts, such as CsCl and NaBr. Dense sucrose solutions are highly viscous, requiring prolonged centrifuge run times. Cesium chloride is expensive, but can be used to make dense solutions of low viscosity. Other materials that have been used to form density gradients include D_2O, NaI, KBr, Cs_2SO_4, RbCl, potassium tartrate, and ficoll, a polysucrose of high molecular weight. Gradients may be formed by layering several solutions

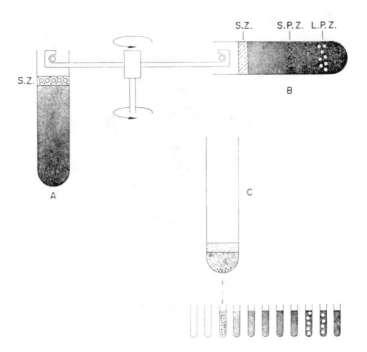

FIGURE 3.4   Rate zonal centrifugation in a swinging bucket rotor. The gradient and sample layer are introduced at rest (A). The tubes are accelerated to a horizontal position (B) and centrifuged until the desired separation is effected. The rotor is decelerated to rest, at which time the gradient and the separated zones (sample zone, S.Z.; small particle zone, S.P.Z.; and large particle zone, L.P.Z.) are recovered (C), usually through a small hole punctured in the bottom of the tube. (Courtesy: National Cancer Institute.)

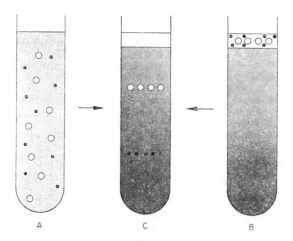

FIGURE 3.5   Isopycnic-zonal centrifugation. The particles to be banded at first may be uniformly distributed in a homogeneous suspending medium (A) and then separated on the basis of buoyant density as a gradient is formed in the centrifugal field (C), or the particles may be layered over a preformed gradient (B) and centrifuged to equilibrium (C). (Courtesy: National Cancer Institute.)

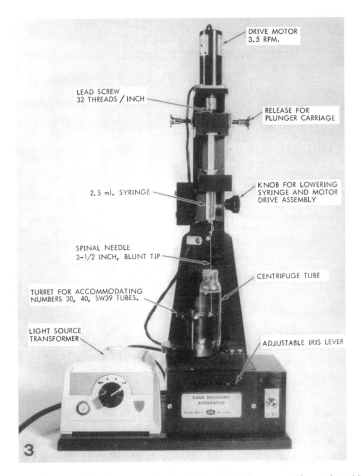

FIGURE 3.6 Apparatus for recovering isopycnically banded particles from centrifuge tubes. (Courtesy: National Cancer Institute.)

of decreasing density, from pipette by hand, and letting the system diffuse to form a smooth gradient. Usually, this procedure takes overnight. Density gradient-forming apparatuses are also used.[146]

The virus preparation is layered on top of the density gradient-containing solution. The tubes are placed in the rotor and centrifuged for the approximate length of time to move the virus zone to the isopycnic position. Fractionation of the centrifuge tubes is accomplished readily if the virus particles are large enough and present sufficient concentration to form a light scattering zone. The virus band may be removed by inserting a syringe with needle at the lower edge of the band and slowly removing the contents. A ratchet and wheel device may be used to move the syringe and needle into the band position (Figure 3.6). Alternatively, the band may be removed by fractionation of the tube contents followed by analysis, using ultraviolet light, infectivity, electron microscopy, or some other revealing indication of the virus particle.

3.1.1.3 Centrifugation with Zonal Rotors[147]

The zonal centrifuge is a unique Oak Ridge invention.[148] Zonal rotor centrifugation was brought into practical utilization during World War II. The U.S. Atomic Energy Commission's Oak Ridge installation was originally organized to achieve certain difficult separations on a large scale. These include the separation of the isotopes of uranium by the electromagnetic process and by gaseous diffusion; the separation of plutonium from fission products and the fractionation of the fission products themselves;

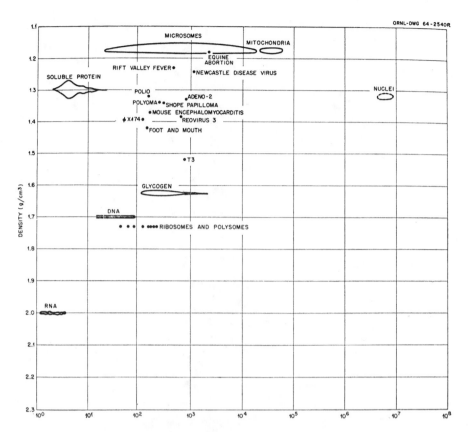

FIGURE 3.7 Sedimentation coefficients and banding densities of cell components and viruses. Most virus particles fall in an area, the "virus window," essentially free of cell constituents. (Courtesy: National Cancer Institute.)

the separation of experimental quantities of all the major stable isotopes of all the naturally occurring elements of the atomic table.[147]

The development of zonal rotors into biological areas of interest was initiated by Anderson and documented in a valuable monograph where design, theory, and application are described. In designing zonal rotors, the following objectives were followed: (1) attaining ideal sedimentation and maximum resolution in sector-shaped compartments, (2) rapid gradient formation in the rotor with minimal stirring or convection, (3) sharp starting or sample zones, (4) high rotational speed, (5) large capacity, and (6) rapid recovery of the gradient after centrifugation without loss of resolution. A number of zonal rotors have been designed to fulfill these requirements.[149]

Anderson analyzed macromolecules on the basis of their sedimentation coefficient, s, and density ρ. The observed distribution of these macromolecules demonstrated that viruses fall on an s-ρ plot in a region unoccupied, generally, by other types of molecules and cellular particulates. This region was termed the "virus window."[150] From Figure 3.7, it can be noted that this area extends from about 100 to 2000 S, and from 1.2 to 1.6 density units. The "virus window" may be used as a guide for approximating the conditions for the isolation and purification of viruses. On the basis of sedimentation coefficient and buoyant densities, the centrifuge may be used to achieve the isolation and purification of viruses and most of the unwanted cellular materials can be eliminated by the selection of proper conditions of centrifugation.

Zonal centrifuge rotors combine the analytical features of density-gradient centrifugation with the high capacity of differential centrifugation. Such assets provide the researcher with great facility in attaining large virus samples of high purity. In this consideration, such features exceed the capabilities of other centrifuge rotors.

FIGURE 3.8 Disassembled B-IV rotor with bearing and seal. (a) End cap; (b) seal; (c) bearing and damper; (d) bottom end cap; (e) rotor chamber; and (f) core. (Courtesy: National Cancer Institute.)

The zonal rotor is a relatively large hollow vessel capable of withstanding the high pressures arising from greater rotational speeds (Figures 3.8 and 3.9). The suspension to be purified nearly fills the entire interior, obviating the use of centrifuge tubes. Loading and unloading the rotor may be performed while it is at rest or spinning at a few thousand revolutions per minute. Initially, a density gradient is introduced, filling the rotor. The sample is layered as a zone on top of the gradient (Figure 3.10). Next, the rotor is disconnected from the feed lines, the vacuum chamber closed, and the rotor accelerated to a high speed. After an appropriate time, the rotor is decelerated back to a low speed, the chamber opened, and the discharge lines connected (Figure 3.11). The gradient is removed from the rotor while still spinning, and the solution is fractionated into a number of samples. Some zonal rotors spin without disconnecting the feed lines, as a bearing assembly attached to the rotor's spindle permits this feature.

The capacities of various zonal rotors vary, but the volumes are usually in excess of 650 ml. The sample size is larger by orders of magnitude than that used in conventional rotors. A welcome feature of zonal rotors is that convection disturbances are absent or minimal, owing in large measure to the small centrifugal field generated by the low rotational speed used during the loading procedure. The combination of high concentration and large volume allows rate zonal rotor methods to be carried out on a preparative scale. In addition, the superior resolving power of the density gradient techniques allows for separations that are far better than may be achieved employing differential centrifugation alone.

Viruses that are damaged in pelleting can be easily freed of this defect by employing zonal rotors. The loading of the sample and unloading of the resulting bands is a very gentle process not realized with other rotor procedures. Zonal rotors have been employed in virus containment operations, where sterility measures are required (Figure 3.12). The purity of vaccines obtained by zonal rotor centrifugation exceeds that from other commonly employed techniques.[151] The use of CsCl and Cs$_2$SO in such large capacity operations may at first preclude their use because of high cost. However, a recovery system was developed to concentrate and purify these gradient-forming materials.[152]

Zonal rotors have provided large yields of insect viruses used in the field as biological insecticides.[153] For such trials, the K-X rotor (Figure 3.13), designed for isolation of micrometer-sized particles, was used to prepare polyhedral inclusion bodies (PIBs) (Figure 3.14) in quantities of 6×10^{13} per run from diseased larvae of insect species (Figure 3.15). The PIBs contain the virus particles, which infect the pest insect host upon ingestion of leaves sprayed with PIBs (Figure 3.16).

FIGURE 3.9   Rotor assembled outside centrifuge, showing damper, bearing, and all lines attached. (Courtesy: National Cancer Institute.)

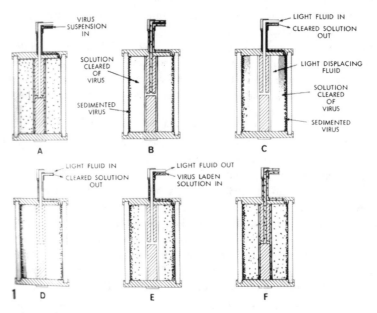

FIGURE 3.10   Batch technique for sedimentation of virus particles using B-II or B-IV zonal centrifuge rotors. (A) Rotor is filled with virus suspension pumped in through edge line; (B) particles are centrifuged to wall; (C) particle-free suspending medium being displaced out of the rotor through edge line by lighter fluid pumped in through core; (D) completion of removal of cleared suspending medium; (E) rotor refilled with virus suspension pumped in through edge line; light fluid displaced out through core; (F) rotor completely filled with virus suspension ready for repetition of the sedimentation cycle. (Courtesy: National Cancer Institute.)

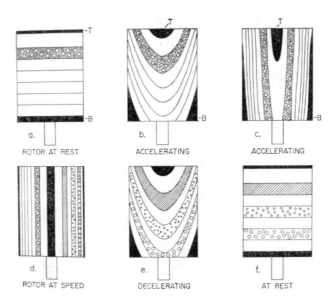

FIGURE 3.11 Schematic diagram of reograd rotor system. (a) Rotor is filled at rest with density gradient and sample layer. To indicate extremes of zone deformation, a thin upper layer, T, and a bottom layer, B, are also indicated. (b) During acceleration, each zone forms a paraboloid of revolutions about the axis. Note T and B. (c) Near operating speed, the zones approach a vertical orientation. (d) At a sufficiently high speed, the zones become nearly vertical. Separation of particle zones is shown at right. (e) During deceleration zones again form paraboloids of revolution. (f) At rest, various zones may be recovered by draining rotor contents out the bottom, or displacing the gradient through the top. (Courtesy: National Cancer Institute.)

FIGURE 3.12 Control room for remote operation of zonal centrifuge. (Courtesy: National Cancer Institute.)

FIGURE 3.13   A K 10 zonal rotor being lowered into the centrifuge chamber. (Courtesy: U.S. Department of Agriculture, Forest Service.)

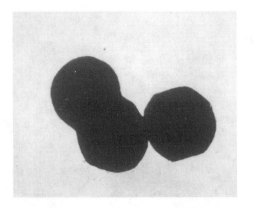

FIGURE 3.14   Viral inclusion bodies purified with the K 10 zonal rotor. (Courtesy: U.S. Department of Agriculture, Forest Service.)

3.1.2   Sedimentation Field Flow Fractionation

The technique of sedimentation field flow fractionation (SdFFF) was first described in 1966 by Giddings.[154] It is used for separating and characterizing both small particles and macromolecules, usually in the size range under 1 μm.

SdFFF is a high-resolution separation technique[155,156] and is a member of a family of field flow fractionation (FFF) methods. All of the FFF methods achieve separation through a common principle

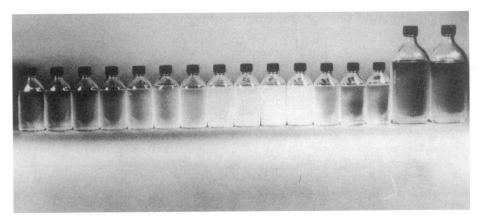

FIGURE 3.15 Harvest of inclusion bodies purified with the K 10 zonal rotor. The four bottles in the center having a whitish appearance are highly purified fractions of inclusion bodies (see Figure 3.14). (Courtesy: U.S. Department of Agriculture, Forest Service.)

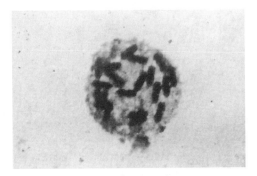

FIGURE 3.16 Chemically digested inclusion bodies revealing virus rod particles. (Courtesy: U.S. Department of Agriculture, Forest Service.)

and can be considered as one-phase chromatographic methods. They are similar to chromatography in that separation is achieved by differential elution of the sample components as they pass through a column or channel. However, in FFF there is no stationary phase or solid support. The sample is injected into a very thin, open channel with a continuous flowing carrier liquid, and components initially are uniformly distributed across the channel. Generation of laminar flow within this narrow channel results in a characteristic parabolic flow profile with maximal flow rate at the center of the channel and flow decreasing to near zero at the channel walls.

An external force field is applied perpendicular to the direction of the flow. This field, which can be temperature gradient, as in thermal FFF, electric potential, as in electrical FFF, hydraulic pressure, as in cross-flow FFF, or centrifugal force, as in sedimentation FFF, causes redistribution of the sample components, with larger or more dense particles forced into slower streams near the channel wall. Smaller or less dense particles are forced into faster flow streams further from the wall. Because of the forces and dimensions of the SdFFF apparatus (Figure 3.17), the particles rapidly achieve sedimentation equilibrium. From back-diffusion against the sedimentation force, the particles do not simply deposit on the channel wall but have a logarithmic concentration distribution near the wall.

As the particles flow down the tube they are separated axially as well as radially, and thus can be collected as preparative fractions. Because the separation is by sedimentation, it relies upon both the viscous resistance, a function of the radius, and the relative densities of the particles. Thus, two particles of the same size but of different composition can be separated.

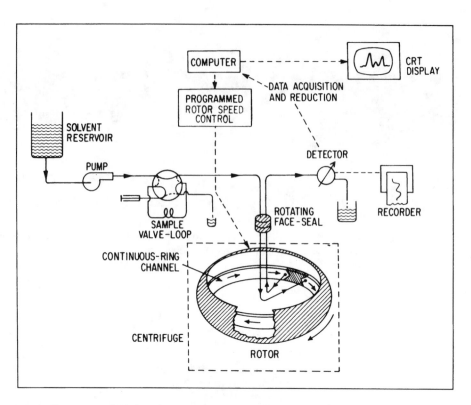

Figure 3.17 Sedimentation field flow fractionation equipment schematic (refer to text). (Courtesy: Electronic Instrumentation and Technology, Sterling, VA.)

The normal elution sequence is one in which small macromolecules elute ahead of large macromolecules. This is the inverse of the elution order found with gel permeation chromatography.[157]

Of the various subtechniques (field types) employed in FFF the one that has proven most adaptable to the physical characterization of virus particles is sedimentation FFF.[158] Viruses are readily analyzed and can be separated from other virus types or from macromolecular contaminants. Viral aggregates, singlets, doublets, and so forth, of the gypsy moth nuclear polyhedrosis virus were easily separated by SdFFF.[159] The large *Paramecium bursaria* chlorella virus (PBCV-1) was physically characterized by this technique.[160] Arner and Kirkland discuss other viruses that have been fractionated by SdFFF, including T2, QB, P22, fd, and T4D.[155]

4

Crystallography

During the 19th and early 20th century a number of protein crystals were observed in animal and plant cells under the light microscope. An early report by Hünefeld in 1840 described hemoglobin crystals forming in a drop of earthworm blood placed between glass plates.[161] Many other protein crystals had been obtained by 1900 and a valuable survey of protein crystals up to that period was made by Schultz.[162] By 1920, Osborne had crystallized a whole series of globulins from plant seeds.[163] Crystallization of several enzymes followed in the 1920s: urease by Sumner,[164] and pepsin, trypsin, and chymotrypsin by Northrop and Kunitz.[165]

Following Stanley's report on the crystallization of tobacco mosaic virus (TMV) in 1935, the ground was broken to study viruses in crystalline form. The first of such studies came from Britain and cast doubt on the TMV crystals Stanley prepared. Bawden and coworkers pointed out that a significant portion of TMV was RNA (6%), which Stanley had not mentioned in referring to his crystals as protein. The pioneer crystallographer, Bernal (Figure 4.1), noted that such crystals were not truly crystalline, lacking a third dimension. Bernal and colleagues referred to Stanley's TMV crystals as paracrystalline.[166,167]

Crystallographic measurements became a major task for elucidating molecular structure following the discovery of X-rays by Roentgen in 1895. The German physicist von Laue was interested in ascertaining the wavelength of X-rays. However, no commercial grating was ruled finely to diffract X-rays, as could be done with ordinary light. It occurred to Laue that crystals consisting of layers of atoms spaced regularly but much more clearly than any manufactured grating would suffice. A beam of X-rays hitting the crystal should then be diffracted as light would be by an ordinary grating. The first X-ray diffraction pattern of any crystal, copper sulfate, was obtained by von Laue, Friedrich and Knipping in 1912. The patterns were too complicated to understand and crystals of zinc blende were analyzed. From such experiments it was possible to correlate the cubic crystal symmetry with that of the diffraction pattern.[168]

This was the starting point of subsequent work on X-rays and crystal structure. These experiments had a twofold result. First, it offered a method of measuring the wavelength of X-rays by using a crystal of known structure and measuring the amount of diffraction. This procedure was investigated by the father-son team of W. H. and W. L. Bragg.[169] Together they worked out the mathematical details leading to the calculation of the wavelength of X-rays and the deduction of certain properties of the crystal from the diffraction pattern.

The second result of the Laue team experiment was that by using X-rays of known wavelength it was possible to study the atomic structure of crystals, where such structure was unknown.[170] In this connection, W.L. Bragg (Figure 2.11) solved the first crystal structures by measuring and comparing the diffraction effects from sodium chloride and potassium chloride.[171] The X-ray diffraction procedure would find great utilization in the case of giant molecules including proteins, nucleic acids, and viruses.

An account of the historical development of X-ray diffraction during the 1920s is given by Mark.[172] Following the work of Laue and coworkers and the Braggs, Nishikawa and Ono in 1913 irradiated a number of fibers, including natural silk, with X-rays. Their work, a pioneering effort, resulted in the recognition that some kind of molecular order was present in these materials. In 1920, Herzog and Jancke obtained "fiber diagrams" of cellulose and silk, concluding that a "crystalline" component was present in the fibers. Polanyi, in 1921, evaluated the fiber diagrams of cellulose and determined its crystallographic basis cell. In 1923, Brill determined the basis cell in silk fibroin, and this work was extended by Meyer and Mark who, in 1928, proposed a chain structure for silk fibroin.

FIGURE 4.1   J.D. Bernal (1901–1971). (Courtesy: Department of Special Collections, Van Pelt-Dietrich Library Center, University of Pennsylvania.)

Arthur Huchinson, a professor of mineralogy at Cambridge University and a forerunner in the teaching of crystallography, was chiefly responsible for the inclusion of this discipline in the University's curriculum in 1927.[173] William T. Astbury and J. D. Bernal, pioneering crystallographers, were taught crystal physics, as undergraduates, by Hutchinson.

Astbury made progress with regard to keratin, myosin, epidermin, and fibrinogen, the k-m-e-f group of proteins. With coworkers Woods and Street he found that the keratin molecule is folded in the natural state, alpha keratin, and unfolded in the stretched state, beta keratin.[174,175] This work would serve as an impetus for Pauling, with Corey and Branson, to extend these studies and led to the alpha helix model for proteins.

In 1934, Bernal and Crowfoot (Figure 4.2) reported the first X-ray diffraction pattern of a protein single crystal, pepsin, in its mother liquor.[176] This was an important advance in the examination of proteins by X-rays. In Bernal's words:[177]

> Until 1934, our only knowledge of the molecular structure of the proteins was derived from the X-ray studies of Meyer, Herzog and Astbury on protein fibers such as silk, collagen, and hair. All of these showed a polymerized chain structure similar to cellulose but presumably of a polypeptide character. Attempts had also been made to investigate crystalline proteins by X-ray methods but always without success. This we now see was due to the fact that the earlier investigators used crystal powders, a useless method with such large (unit) cells, and also worked with dry material in which the crystal structure of the protein had collapsed.
>
> In the spring of 1934, I was fortunate to obtain crystals of crystallized pepsin prepared by Philpot, by Northrop's method, and, foreseeing the above difficulties, I examined a single crystal bathed in its mother liquor. A good diffraction pattern was obtained and the main outlines of the crystal structure were made out. The unit cell is relatively enormous — 461 × 67 × 67 Å — of molecular weight 36,000, independently confirming Svedberg's values.

X-ray diffraction determinations of the volumes of the unit cells of crystalline proteins are useful in giving a figure which must be an integral multiple of the molecular weight — plus the weight of any water or other substances crystallographically or chemically combined with the protein.[178]

FIGURE 4.2   Dorothy Crowfoot Hodgkin with M.A. Rafferty and A. Rich (1971). (Courtesy: Cold Spring Harbor Laboratory Archives.)

In collaboration with Bawden and Pirie, Bernal and Fankuchen studied TMV with X-rays. Bernal and Fankuchen were able to derive the shape and approximate size of the virus particle. These observations, made before the advent of the electron microscope, indicated the virus particles as being rod shaped, 150 Å across, and about 1500 Å in length. The studies also indicated structure within the virus particles.

Crystals of tobacco necrosis virus were studied by Crowfoot and Schmidt in 1945;[179] turnip yellow mosaic virus with nucleic acid and without it, by Bernal and Carlisle in 1948;[180] and tomato bushy stunt virus in 1948 by Carlisle and Dornberger.[181]

Protein crystallography continued to reap benefits in the Laboratory of Molecular Biology at Cambridge University. There, hemoglobin and myoglobin were studied by X-ray crystallography by Perutz and Kendrew, respectively. It occurred to Perutz that the heavier an atom the more efficiently it would diffract X-rays. Perutz experimented with the attachment of mercury, in the form of *para*-mercuribenzoate, to hemoglobin. The diffraction picture of hemoglobin was altered significantly, and facilitated the process of assigning atom positions from the diffraction pattern.[182] These analyses were greatly aided through the use of high-speed computers, pioneered by Kendrew in his work on myoglobin, the first globular protein to have its structure resolved crystallographically.[183]

From X-ray diagrams obtained by Franklin and Gosling[184] and by Wilkins and Randall,[185] DNA extracted from the cell was shown to have two forms. The B form is present when DNA is fully hydrated, as it is *in vivo*. The A form results when DNA is dehydrated. One of the earliest X-ray diagrams of DNA was that of Astbury and Bell in 1938 on DNA extracted by Hammarsten. This preparation was a mixture of A and B forms and among the conclusions reached were the following: a meridional arc exists at 3.34 Å and an equatorial spot at 16.2 Å. The meridional arc was attributed to the purine and pyrimidine bases piled one upon another like a pile of pennies, and at right angle to the fiber axis.[105]

From a consideration of the hierarchical orders of protein structure, the following general observations are made: the primary structure is obtained by direct determination of the amino acid sequence from the protein or indirectly, from the nucleotide sequence of the corresponding gene or complementary DNA (cDNA).[186]

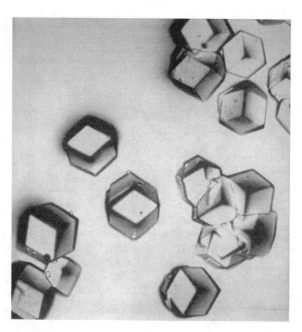

FIGURE 4.3   Crystalline bushy stunt virus × 224. (From Stanley, W.M., *J. Biol. Chem.*, 135, 437, 1940. With permission.)

The importance of the primary structure in X-ray diffraction studies was recognized early. Bernal in 1931 noted:[187] "…a knowledge of the crystal structure of the amino acids is essential for the interpretation of the X-ray photographs of animal materials: silk fibroin, keratin, collagen, proteins, etc., which have been studied by this method for the first time in the last few years."

Branden and Tooze note that the quaternary structure of large proteins or conjugated nucleoproteins, such as viruses, can be determined by electron microscopy. Usually the resolution is low, but with refinement and residual noise elimination, the shape of the protein subunits may be revealed. In order to obtain the secondary and tertiary levels of structure in which detailed information on the arrangement of atoms within a protein are required, X-ray crystallography and nuclear magnetic resonance (NMR) methods are used. The latter method works well only with small protein molecules but crystallization is not a prerequisite.[188]

Crystals are three-dimensional objects and the first prerequisite for resolving the structure of a protein by X-ray crystallography is a well-ordered crystal that will diffract X-rays strongly (Figure 4.3). Crystals may be grown from a variety of conditions, including supersaturated solutions, supercooled melts and vapors. The formation of a crystal may be considered in two steps. The first step, nucleation, is the coming together of a few atoms to form a three-dimensional periodic array — the nucleus — which will show faces, although it is only a few unit cells in size. The second step is the growth of the nucleus into a crystal. The nucleus attracts other atoms, and they take up positions on its faces in accordance with its three-dimensional periodicity. This results in new lattice planes being formed. The growth of the nucleus, and then of the crystal, is marked by a parallel displacement of its faces. The rate of displacement is called the rate of crystal growth, and is a characteristic, anisotropic property of a crystal.[189]

A commonly used procedure for making protein crystals is the hanging drop method (Figure 4.4). A drop of protein solution is brought very gradually to supersaturation by loss of water from the droplet to a small container that contains salt or polyethylene glycol solution. Equilibrium between the drop and the container is slowly reached through vapor diffusion, and the salt concentration in the drop is increased. If other conditions such as pH and temperature are right, protein crystals will occur in the drop. Ammonium sulfate is the single most effective crystallization reagent. Polyethylene glycol is also routinely used to enhance crystallization.

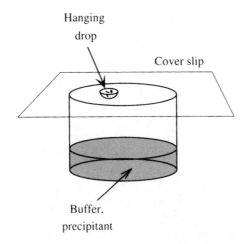

FIGURE 4.4 Growing crystals by the hanging-drop method. The droplet hanging under the cover slip contains buffer, precipitant, and protein. (From Rhodes, G., *Crystallography Made Crystal Clear, A Guide for Macromolecular Models,* Academic Press, New York, 1993. With permission.)

In 1992, experiments on the effect of gravity on crystal growth were conducted on a U.S. Space Shuttle mission designated International Microgravity Laboratory–1 (1ML-1), flown January 22–29, 1992. Day and McPherson compared their observations on proteins and virus crystal growth on earth and under microgravity conditions.[190,191] Unusually impressive results were achieved for satellite tobacco mosaic virus (STMV). STMV crystals grown in microgravity by liquid-liquid diffusion were more than 10-fold greater in total volume than STMV crystals previously grown in the laboratory. Under microgravity conditions constant temperature experiments produced crystals of uniform habit and almost perfect shape. X-ray diffraction data collected from STMV crystals under microgravity conditions demonstrated a substantial improvement in diffraction quality over the entire resolution range when compared to data from crystals grown on earth. The diffraction pattern for the STMV crystals grown in space extended to 1.8 Å resolution, whereas the best crystals that were ever grown under conditions of Earth's gravity produced data limited to 2.3 Å resolution.

STMV is a very small icosahedral virus having a total weight of 2×10^6.[192] Figure 4.5 is a crystal of STMV grown under microgravity conditions.

In entering the crystalline state from solution, individual molecules of the substance adopt one or only a few orientations. The resulting crystal is an orderly three-dimensional array of molecules, held together by noncovalent interactions.

The crystal is divided into identical repetitive sections, unit cells. The array of points at the corners or vertices of unit cells is called the lattice. The unit cell is the smallest and simplest volume element that is completely representative of the whole crystal (Figure 4.6).

The unit cell may contain one or more than one molecule, but the number of molecules on each unit cell is always the same for all the unit cells of a single crystal. However, a number of different crystal forms may exist for the same protein, and the number of molecules per unit cell may vary between different forms of the same protein. In terms of crystal orientation, there are 7 systems comprising 47 different crystal forms.[189]

In the crystallographic process a narrow and parallel beam of X-rays is directed onto a crystal. The X-ray source is usually a high-voltage tube in which a metal plate, the anode, is bombarded with accelerating electrons causing the emission of X-rays. More powerful X-ray beams are produced in synchrotron storage rings. If the wavelength of the X-ray is specific, it is monochromatic, and if of variable wavelength, it is polychromatic. Both types are used to determine structure by X-ray crystallography.

The primary beam of X-rays striking a crystal has two results. Most of the X-rays travel straight through the crystal. Some, however, interact with the electron cloud of each atom, causing the electrons to oscillate.

FIGURE 4.5   Satellite tobacco mosaic virus crystal grown in microgravity — NASA Space Shuttle. (Courtesy: Dr. Alexander M. McPherson, University of California, Riverside.)

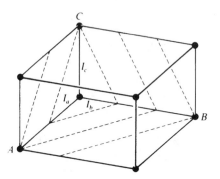

FIGURE 4.6   A crystallographic unit cell. The basis of the description of the structure of a crystal is the unit structure, the unit cell. It is the smallest group of atoms, ions, or molecules, whose repetition at regular intervals, in three dimensions, produces the lattice of a given crystal. There are seven basic types of unit cells which result in the seven crystal systems. (From Pauling, L., *General Chemistry*, Dover Publications, New York, 1988. With permission.)

The oscillating electrons produce new X-rays, which are emitted in almost all directions. This scattering diffraction of X-rays interfere with one another. In most cases, X-rays colliding from different directions cancel each other. But X-rays from certain directions will add together to produce diffracted beams of radiation that can be recorded as a pattern of diffracted spots (Figure 4.7).

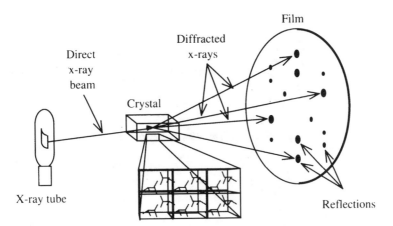

FIGURE 4.7   Crystallographic data collection. The crystal diffracts the source beam into many discrete beams, each of which produces a distinct spot (reflection) on the film. The positions and intensities of these reflections contain the information needed to determine molecular structures. (From Rhodes, G., *Crystallography Made Crystal Clear, A Guide for Macromolecular Models*, Academic Press, New York, 1993. With permission.)

The diffraction spots are recorded either on film, the original method, or by an electronic detector. Use of the latter method, whereby signals are fed directly into a computer, has significantly reduced the time required to collect and measure diffraction data.

The rules for diffraction of X-rays were established by Sir Lawrence Bragg of Cambridge University in 1913.[171] The diffraction by a crystal can be regarded as the reflection of the primary beam by sets of parallel planes through the unit cells of the crystal (Figure 4.8). X-rays that are reflected from adjacent planes travel different distances, and diffraction only occurs when the difference in distance is equal to the wavelength of the X-ray beam (Figure 4.9). This distance is dependent upon the reflection angle, which is equal to the angle between the primary beam and the planes. The relationship between the reflection angle, θ, the distance between the planes, d, and the wavelength, λ, is set by Bragg's law: $2d \sin \theta = \lambda$. This equation is used to determine the size of the unit cell.

In measuring the unit cell, the crystal is oriented in the beam of X-rays so that reflection is obtained from the specific set of planes in which any two adjacent planes are separated by the length of one of the unit cell's three axes. This distance, d, is then equal to λ ($2\sin\theta$). The wavelength λ of the beam is known in monochromatic radiation. The reflection angle θ can be calculated from the distance (r) between the diffracted spot on the film, and the position where the primary beam hits the film. The tangent of the angle $2\theta = r/A$, where A is the distance between the crystal and film that can be measured on the experimental equipment, and r can be measured on the film. The value of θ is then obtained. The crystal is then reoriented, and the entire procedure repeated for the other two axes of the unit cell.

A diffraction pattern reveals the distribution of electrons, or the electron density of the molecules. Electron density reflects the shape of the molecule(s) in a unit cell as a three-dimensional periodic function, p (x,y,z). A graph of the function is called an electron-density map. The map, a contour map, is a blurred image of the molecules in the unit cell. The crystallographer attempts to obtain the mathematic function whose graph most closely resembles the true electron-density map. This task is accomplished through a mathematic operation called the Fourier transform. Its application to electron density calculations was first suggested by Bragg in 1915.[169] The Fourier transform describes the mathematical relationship between an object and its diffraction pattern by describing p (x,y,z) in terms of the amplitude, wavelength, and phase of each reflection or spot on the diffraction pattern. Amplitude is a measure of the strength of the diffracted beam and is proportional to the intensity of the recorded spot. The wavelength is set by the X-ray source for monochromatic radiation. Phase is related to the interference, positive or negative, by the diffracted beam with other beams.

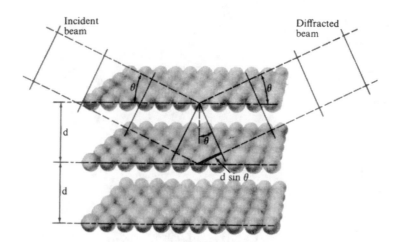

FIGURE 4.8 Diagram illustrating the derivation of the Bragg equation for the diffraction of X-rays by crystals. If the beam of rays incident on a plane of atoms and the scattered beam are in the same vertical plane and at the same angle with the plane, the conditions for reinforcement are satisfied. Bragg formulated the conditions for reinforcement of the beam specularly reflected from one plane of atoms and the beam specularly reflected from another plane of atoms separated from it by the interplanar distance d. The difference in path is equal to $2d \sin \theta$, in which θ is the Bragg angle (the angle between the incident beam and the plane of atoms). In order to have reinforcement, this difference in path $2d \sin \theta$ must be equal to the wavelength λ or an integral multiple of this wavelength — that is the $n\lambda$, in which n is an integer. Thus, the Bragg equation for the diffraction of X-rays is obtained: $n\lambda = 2d \sin \theta$. (From Pauling, L., *General Chemistry*, Dover Publications, New York, 1988. With permission.)

The phase parameter is lost in X-ray experiments. For large molecules, proteins and viruses, the phase problem may be circumvented by recourse to a method, noted above, pioneered by Perutz[182] and Kendrew[183] — multiple isomorphous replacement (MIR). By this strategy heavy atoms, such as mercury or gold, are introduced into the molecules of the sample. These additions make a significant contribution to the diffraction pattern without changing the structure of the molecule or the crystal cell — the crystals remain isomorphous. Heavy metals contain many more electrons than the light atoms in the sample molecule, and hence scatter X-rays more strongly. Following heavy metal substitution, some diffraction spots measurably increase in intensity (positive interference). Others decrease (negative interference), and many show no detectable difference. From intensity differences the positions of the heavy atoms in the crystal unit cell may be deduced. Fourier analysis, calculated with the squared amplitudes of the X-ray spectra as coefficients, define distributions in which the density corresponds with inner atomic vectors in the crystal. From these analyses, referred to as Patterson maps,[193] the atomic arrangement of the heavy atoms is derived.

The amplitude and phase of the heavy metal contribution to the diffracted beam can also be calculated. The following facts then become known: the phase and amplitude of the heavy metals, the amplitude of the sample molecule, the amplitude of the sample plus heavy metals. Thus one phase and three amplitudes are realized. From this information, one can calculate whether the interference of the X-rays scattered by the heavy metals and protein is constructive or destructive. An estimate of the phase of the protein is given by the extent of positive or negative interference and knowledge of the phase of the heavy metal. Two different phase angles result as equally good solutions. To resolve the issue as to which solution is correct, a second heavy metal complex is used, which also gives two possible phase angles. However, in the second case, only one of the angles will have the same value as one of the two previous angles, thus representing the correct phase angle. In practice, many different heavy metal complexes are tried to rule out experimental errors and give reasonably good phase determinations.[194]

The amplitudes and the phases of the diffraction data from protein or virus crystals are used to calculate the electron densities of the repeating unit of the crystal. The electron density map is interpreted as a

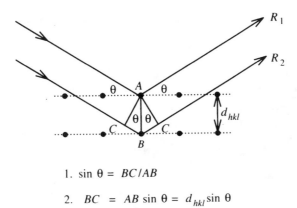

1. $\sin \theta = BC/AB$

2. $BC = AB \sin \theta = d_{hkl} \sin \theta$

FIGURE 4.9 Conditions that produce strong diffracted rays. If the additional distance traveled by the more deeply penetrating ray R_2 is an integral multiple of λ, then rays R_1 and R_2 interfere constructively. (From Rhodes, G., *Crystallography Made Crystal Clear, A Guide for Macromolecular Models*, Academic Press, New York, 1993. With permission.)

polypeptide chain having a particular amino acid sequence. The quality of the map depends on accurate phase angles and on the resolution of the diffraction data. Resolution in turn depends on a well-ordered crystal, which directly influences the quality of the image produced — the diffraction pattern. Resolution is measured in angstrom units (Å). One angstrom equals 0.1 nm. The smaller the resolution number, the higher the resolution and the greater the amount of detail that can be seen on the electron density map.

The map obtained is interpreted by building a model that fits it. For a protein, knowledge of the amino acid sequence is essential in order to accomplish this task. The final model must have certain features: (1) it must be consistent with the image; (2) it must be chemically realistic; (3) it must possess bond lengths, bond angles, and conformational angles, and distances between structural groups that follow established principle of molecular structure and stereochemistry.[195] Refinement of the model involves the investigator working back and forth between two different coordinate systems.

There is an inverse relationship between the spacing of unit cells in the crystalline lattice, the real lattice, and the spacing of reflections in the lattice on a record, such as film, the reciprocal lattice. Crystallographers calculate the dimensions, in angstroms, of the unit cell of the crystalline material for the spacings of the reciprocal lattice on the X-ray film.

The first system is the unit cell-real space, where the position of an atom is described by its coordinates, x,y,z. The second system is the three-dimensional diffraction pattern, reciprocal space, where the position of a reflection is described by indices, hkl. Distances in reciprocal space, expressed in reciprocal angstroms or reciprocal nanometers, are used to judge the potential resolution that the diffraction data can yield. Map calculations and model building, forms of real space refinement of the model, are analyzed together with computerized analyses to improve the agreement of the model with the original intensity data. Since the computations involve comparison of measured and observed structure factor amplitudes (reciprocal space, rather than examinations of maps and models — real space), these methods are referred to as reciprocal space refinement. As shown in Figure 4.10, the crystallographer compares data between real and reciprocal space to mold the model into the best possible representation with the data.[195] As an example of refinement of crystallographic data in reciprocal space, the reader is referred to analyses by Silva and Rossman on southern bean mosaic virus.[196]

Levels of resolution permit the investigator to analyze gradation of structure. At low resolution, 5 Å or higher, the shape of the molecule and sometimes the alpha-helical regions can be obtained. At medium resolution, about 3 Å, it is usually possible to trace the path of the polypeptide chain and to fit a known amino acid sequence in the map. In addition, at 3 Å resolution, gross features of functionally important aspects of a structure usually can be deduced, including the identification of active site residues. At 2 Å

Collecting Diffraction Data

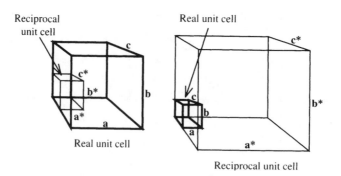

FIGURE 4.10   Reciprocal unit cells of large and small real cells. (From Rhodes, G., *Crystallography Made Crystal Clear, A Guide for Macromolecular Models,* Academic Press, New York, 1993. With permission.)

resolution, details allow the investigator to differentiate similarly structured amino acids, such as leucine and isoleucine. Atoms are viewed as discrete balls of density at 1 Å resolution. The structures of only a few small proteins have been determined to such high resolution.[197]

Fry et al.[198] have reviewed the progress of virus crytallography. The computational requirements were satisfied in the late 1970s, and when combined with the introduction of phase improvement techniques utilizing the virus symmetry,[199,200] the application of crytallography to the massive macromolecular assemblies of viruses became feasible. The determination of the first virus structure, the small RNA plant virus, tomato bushy stunt virus,[201] was reported in 1978. The structures of two other plant viruses followed: southern bean mosaic virus[138,202] and satellite tobacco necrosis virus.[203]

A major focus of attention was a family of animal RNA viruses — the Picornaviridae. The determination of the structures of a human rhinovirus[138,204] and that of a poliovirus[205] became landmarks in macromolecular crystallography. The 1990s has witnessed notable progress and insight, with results on the RNA bacterial virus MS2,[206] the DNA bacterial virus φX174,[207] and two DNA animal viruses, canine parvovirus[208] and SV40.[209]

These structural studies have shown that the arrangement of subunits within a spherical capsid does not adhere strictly to the theory of virus architecture known as quasi-equivalence, advocated by Caspar and Klug (discussed in Chapter 2). As stated, this theory allows 60T subunits to be accommodated in a spherical shell, where T subunits occupy quasi-equivalent environments and T is subject to some exclusion rules. Thus, there is not a single mechanism controlling the assembly of these viruses. It was noted, however, that a striking similarity has been seen in the topology and tertiary structure of the coat proteins leading to a more unified view of the evolution of seemingly rather disparate groups of animal, insect, and plant RNA viruses.[198]

Fry et al.[198] list a number of viruses studied crytallographically to date. For most of these, T = 1 or 3. The Picornaviridae are particularly well studied, with at least one structure known from four different genera. Some present studies are attempting to analyze larger viruses. The structure of SV40 has been reported.[209] Preliminary results have been published for rice dwarf virus,[210] bluetongue virus cores,[211,212] cauliflower mosaic virus,[213] and reovirus type 3 Dearing cores.[214]

Crystallographic studies of viruses are necessarily incomplete. As noted, while the external protein coat or capsid with its symmetrical subunits can be analyzed by crystallography, the nucleic acid genome, residing within the protein shell of the virus, is not seen in X-ray diffraction. Table 4.1 lists several viruses that have been analyzed by X-ray diffraction at various resolution levels.

Crystallographic data on proteins and viruses are commonly entered into data banks. Figure 4.11 presents such data on a strain of poliovirus from one data bank. The availability of data banks is discussed in Chapter 8.

```
                NIST/NASA/CARB BMCD User File Report for VIRUS        03/30/95
Macromolecule Name:                                                   ME#: M177
virus, polio-, type 2/type 1, V510, mouse adapted
     genus: Picornavirus                    tissue:
     species:                               cell:
     common name:                           organelle:
  Total Molecular Weight:     5800000 Total No.  Subunits:      240
              alpha   Mwt:     33000          No. Subunits:      60
               beta   Mwt:     30000          No. Subunits:      60
              gamma   Mwt:     26000          No. Subunits:      60
              delta   Mwt:      7500          No. Subunits:      60
Remarks:
V510 differs from its type 1 parent only in the replacement of a single external
loop with the residues from a type 2 poliovirus.


Total Number of Crystal Entries:    1



Crystal entry CE#: C1T7
   a        b        c      alpha   beta   gamma   Z Space Group       CS
323.30   358.50   380.50   90.00   90.00   90.00   2 P2⟨1⟩2⟨1⟩2    Orthorhombic


Method(s) used for the crystallization:
          microdialysis
Macromolecule concentration:    10.000 mg/ml
Temperature:    4.0 degrees C
pH: 7.000
Chemical additions to the crystal growth medium:
          sodium chloride                       0.2500   M
          PIPES                                 0.0100   M
          magnesium chloride                    0.0050   M
          calcium chloride                      0.0010   M
Diffraction limit:    2.600 Angstroms
Comments:
The virus was dialyzed against progressively lower concentrations of sodium chloride
in the same buffer. Crystals generally appeared at about 0.1 M sodium chloride.


References:

R24A:
Yeates, TO; Jacobson, DH; Martin, A; Wychowski, C; Girard, M; Filman, DJ;
Hogle, JM (1991) EMBO J, 10(9), 2331–2341.
"Three-dimensional structure of a mouse-adapted type 2/type 1 poliovirus chimera."
```

FIGURE 4.11 Crystallographic data on polio virus available from the NIST/NASA/CARB/BMCD Database.

TABLE 4.1 Some Representative Viruses Analyzed by X-ray Crystallography at High Resolution

Virus	Resolution (Å)
Bean-pod mottle[215]	3.0
Black beetle[216]	3.0
Foot-and-mouth disease[217]	2.9
Mengo[218]	3.0
Polio type 1 (Mahoney strain)[205]	2.9
Satellite tobacco mosaic[219]	2.9
Satellite tobacco necrosis[220]	2.5
Sesbania mosaic[221]	4.7
Southern bean mosaic[202]	2.8
Tomato bushy stunt[201]	2.9
Turnip crinkle[222]	3.2

5

Molecular Weight Determinations from Sedimentation Experiments

5.1 Analytical Ultracentrifugation

Analytical ultracentrifugation is a versatile and accurate means for obtaining the molecular weight of a macromolecule. It is unsurpassed for the direct measurement of molecular weights of solutes in the native state and as they exist in solution and, unlike other methods, it is not dependent on standards.[223]

Two distinct types of procedure are used in analytical ultracentrifugation: (1) sedimentation velocity, in which the velocity of sedimentation of a component of the solution is measured, and (2) sedimentation equilibrium, in which the redistribution of molecules is determined at equilibrium. From such procedures Svedberg derived two fundamental equations to calculate the molecular weight of the sedimenting particles.[224] These equations are discussed below.

In early studies with the ultracentrifuge it was realized that deviations from ideal conditions are often very large, even at low concentrations. Large errors in reported molecular weight values resulted. The reason for such deviations was attributed to the high molecular weight of the solute. To counter such errors it became common practice to extrapolate the experimental results to infinite dilution — zero concentration.

5.1.1 Instrumentation

Many of the experiments performed with the analytical ultracentrifuge were done on the venerable Beckman Model E (Figure 5.1). In recent years the Beckman Model E has been replaced by the Beckman Optima series,[225] which serves as both preparative and analytical ultracentrifuges (Figure 5.2). The following discussion is concerned with both types of analytical ultracentrifuge.

5.1.1.1 Rotors

The rotor of an analytical ultracentrifuge is capable of speeds up to 60,000 RPM under controlled temperature in an evacuated chamber. The rotor is so constructed that passage of light through the spinning sample monitors the progress of the run by recording the movement of the macrosolute particles (Figure 5.3).

The Optima Analytical Ultracentrifuges utilize rotors containing several holes. One of the holes is required for the counterbalance, with its reference holes that provide calibration of radial distance. The remaining positions in the rotor are available for sample cells, allowing multiple samples to be analyzed in a single experiment (Figure 5.4).

FIGURE 5.1   The Model E Analytical Ultracentrifuge. (Courtesy: Beckman Instruments, Palo Alto, CA.)

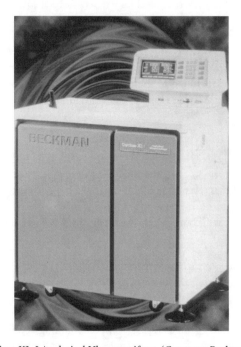

FIGURE 5.2   The Model Optima XL-I Analytical Ultracentrifuge. (Courtesy: Beckman Instruments, Palo Alto, CA.)

FIGURE 5.3 Analytical rotor for the Model E Analytical Ultracentrifuge. (Courtesy: Beckman Instruments, Palo Alto, CA.)

FIGURE 5.4 Rotors for the Optimal Series of analytical ultracentrifuges. Top: Rotor for the XL-A Analytical Ultracentrifuge. Bottom: Rotor for the Optima XL-I Analytical Ultracentrifuge. (Courtesy: Beckman Instruments, Palo Alto, CA.)

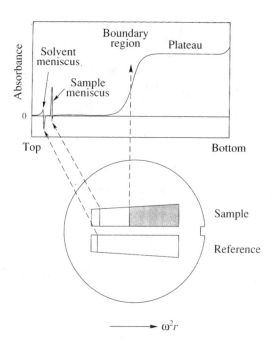

FIGURE 5.5   Relationship in a sector-shaped cell of the sedimenting regions of solution. The figure shows a double sector centerpiece. The sample solution is placed in one sector, and a sample of the solvent in dialysis equilibrium with the sample is placed in the reference sector. The reference sector is usually filled slightly more than the sample sector, so that the reference meniscus does not obscure the sample profile. (Courtesy: Beckman Instruments, Palo Alto, CA.)

5.1.1.2   Cells and Avoidance of Convection

Convection was a problem recognized in early sedimentation studies carried out by Svedberg and coworkers.[224] In situations where the action of gravity is of concern, convection may be defined as the circulatory motion that occurs in a solution at a nonuniform temperature resulting from density variations. In ultracentrifuge runs, convective flow produces the following effects. For an area of greater density than its immediate surroundings, transport of solute readily proceeds toward the bottom of the cell. Conversely, for an area of lesser density than its immediate surroundings, transport of solute is in the direction of the meniscus. As a result, measurements on optical patterns obtained in such cases are doubtful.

The conditions for convection-free sedimentation were studied by Svedberg and Rinde.[226] It was found that the sample cell must be sector-shaped (Figure 5.5), completely enclosed, and of not too large dimensions. In sedimentation velocity runs, the sector shape is critical. The apex of the sector is at the center of rotation in order to allow the particles of solute to sediment unobstructed along radii.

Sector-shaped cells are convection-free barring inverse thermal gradients or the misalignment of the cell. In density gradient runs, the presence of a gradient of appreciable density counters any serious convective disturbances, permitting the use of angle-head rotors (see Figure 3.2).

In sedimentation equilibrium experiments the column of solution may have any shape. The relatively low speeds used in such runs makes no restriction on cell shape.

For sedimentation velocity runs the sample is usually contained within a sector-shaped cavity lying between two thick windows of optical grade quartz or sapphire. The cavity is produced in a centerpiece of aluminum alloy, reinforced epoxy, or a polymer known as Kel-F.*

---

* A registered trademark of the 3M Company.

Double sector centerpieces for the Optima series centrifuges are available with optical lengths of 3 and 12 mm. These cells allow the user to take account of absorbing components in the solvent, and to correct for the redistribution of solvent components, particularly at high *g* values. A sample of the solution is placed in one sector, and a sample of the solvent, in dialysis equilibrium with the solution, is placed in the second sector (Figure 5.5). The optical system measures the difference in absorbance between the sample and reference sectors. Double sector cells also facilitate measurements of differences in sedimentation coefficients and diffusion coefficients.

In equilibrium experiments, the time required to attain equilibrium is considerably reduced by employing cells with shorter columns of solution. The distance from the meniscus to the cell bottom is about 1 to 3 mm, rather than the 12 mm for a full sector cell. Three samples may be analyzed at one time utilizing six channel centerpieces. Three channels hold three different samples and the corresponding three channels on the other side hold the respective dialyzed solvents.

Also available are special cells,[227] such as the synthetic boundary forming cell. Use of this cell allows the solvent to be layered over a sample of solution while the cell is spinning at moderately low speed. This type of cell is useful for preparing an artificial sharp boundary for measuring boundary spreading in measurements of diffusion coefficients. The cell also permits the analysis of small molecules, for which the rate of sedimentation is insufficient to produce a sharp boundary that clears the meniscus. The synthetic boundary cell is also used to obtain the initial solute concentration (c_o) discussed below.

The band-forming cell allows the layering of a small volume of solution on top of a supporting density gradient whereby a band is produced for analysis. The use of such a cell allows for density gradient experiments to be conducted in the analytical ultracentrifuge. Previously such experiments were possible only in preparative ultracentrifuges.

The use of the band-forming cell was pioneered by Vinograd and coworkers in their study of macromolecules, including viruses, in self-generating density gradients.[228]

The technique provides estimates of the sedimentation coefficient in a comparatively short time using relatively small amounts of material. Moving molecules are contained within bands or zones, and a complete physical separation of bands will occur given a sufficient difference in the sedimentation coefficients of the components in a mixture. The physical separation will eliminate the Johnston-Ogston effect arising from solute interactions that may occur in boundary sedimentation velocity experiments (see below).

Sedimentation coefficients and molecular weights of proteins have been reported in other studies.[229,230] Recently, band cells in the Beckman Model E and Beckman XL-A analytical ultracentrifuges were used to monitor the stability of the human immunodeficiency virus-1 reverse transcriptase heterodimer.[231,232]

5.1.2 Optical Systems

Three types of optical systems are used in sedimentation studies: absorption, schlieren, and Rayleigh interference (Figure 5.6). Each method has advantages and disadvantages, depending on such factors as solute concentration, sensitivity, and accuracy (Table 5.1).

In absorption optics (UV-visible) concentrations are measured directly in absolute terms without the need for an integration process which is required with schlieren optics, or for the labeling of fringes produced by interference optics. The schlieren optical system is the simplest, most versatile, and most common of the optical systems used in sedimentation velocity analysis. It provides direct viewing of the movement and distribution of macromolecules in a centrifugal field. The interference optical system is the most accurate of the three systems. Many sedimentation experiments, especially sedimentation equilibrium runs, are analyzed by interference optics, whereby concentration information is obtained by interpretation of fringes.

5.1.2.1 Absorption Optics

This system was the first employed by Svedberg and coworkers. It was based on the absorption of light by sedimenting particles. The early absorption optical system was neither convenient to use nor accurate.[233] In

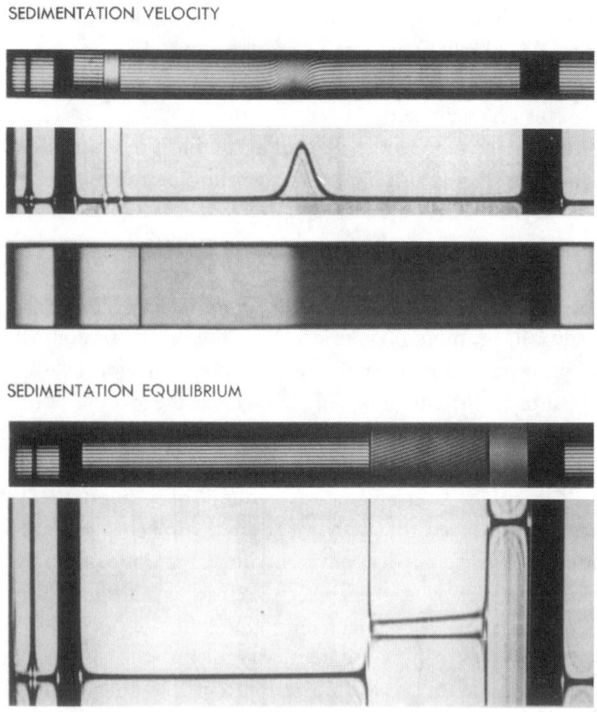

SEDIMENTATION VELOCITY

SEDIMENTATION EQUILIBRIUM

FIGURE 5.6   Optical patterns from analytical ultracentrifuge runs. Sedimentation velocity: (top) Rayleigh interference fringes; (middle) schlieren pattern; (bottom) absorption (ultraviolet). Sedimentation equilibrium: (top) Rayleigh interference fringes; (bottom) schlieren pattern. (From Schachman, H. K., *Ultracentrifugation in Biochemistry*, Academic Press, New York, 1959. With permission.)

the 1950s, considerable refinement resulted in the photoelectric scanner, which is now used for all types of sedimentation studies.

The great benefit of the absorption optical system is its sensitivity. Many biological macromolecules absorb light in the ultraviolet region, and their migration in a centrifugal field can be measured readily with a monochromator. Viruses, nucleic acids, and proteins can be analyzed in solutions containing only a few micrograms per milliliter. Such sensitivity is not achieved in schlieren or interference optics.

The photoelectric scanner rapidly and directly produces plots of concentration (absorbancy) and concentration gradient vs. position of the sedimentation boundary (Figure 5.7). Such data can be automated by coupling the electric pulses from the photomultiplier to computers.

Another feature of the absorption optical system is its capacity for discrimination of the components of sedimenting systems. Such components vary in their absorption properties and are distinguished one from another. One can observe the sedimentation of a specific component by proper selection of the wavelength of light. In multicomponent systems containing interacting components, the solute responsible for the interaction, e.g., urea, can be rendered "invisible" while the change in migration of protein can be analyzed by the appropriate selection of the wavelength of incident light.

The pattern produced by the recording system gives a plot of concentration vs. distance in the cell (Figure 5.8). Derivative patterns, obtained through differentiating circuits, may also be obtained, and yield plots of concentration gradient vs. distance. Such patterns show directly the number, shapes, and position of boundaries in sedimentation velocity runs. Together, the integral and derivative patterns are very useful in detecting the presence of either slowly or rapidly sedimenting species.[233]

TABLE 5.1 Comparison of the Optical Systems Used in Analytical Ultracentrifugation

	Schlieren	Interference	Absorption	
			Photographic UV	Photoelectric Scanner
Most important applications	Sedimentation velocity Studies of heterogeneity Approach to equilibrium	Sedimentation equilibrium Concentration determination	Sedimentation velocity (dilute solutions) Equilibrium banding	Sedimentation equilibrium (dilute solutions) Sedimentation velocity (dilute solutions) Equilibrium banding
Other usual applications	Sedimentation equilibrium Concentration determination	Diffusion studies Sedimentation velocity		Small molecule binding Differential sedimentation
Usual concentration ranges (12 mm cells)	1.0–10 mg/ml	0.5–5 mg/ml	Proteins: 0.1–1.0 mg/ml Nucleic acids: 0.01–0.1 mg/ml	Proteins: 0.05–1.0 mg/ml Nucleic acids: 0.01–0.1 mg/ml
Special advantages	Direct viewing Direct determination of concentration gradients Heterogeneity visualized Variable sensitivity (phase plate angle)	Direct viewing Direct determination of relative concentrations High accuracy	Discrimination among solutes Applicable to very dilute solutions	Baseline provided Equivalent to direct viewing Discrimination among solutes Applicable to very dilute solutions
Particular disadvantages	Salts may interfere Integration required to obtain concentration Patterns sometimes difficult to read accurately	Differentiation required to obtain concentration gradients Sensitive to cell distortion	No direct viewing Inconvenient Sensitive to oil and dirt on optical components	Sensitive to oil and dirt on optical components

From Chervenka, C.H., *A Manual of Methods for the Analytical Ultracentrifuge,* Beckman Instruments, Palo Alto, CA, 1969. With permission.

Compared to interference optics the absorption optical system is less accurate. It is comparable, however, to schlieren optics. For some studies, nucleic acids for example, the concentrations required for optimal use of the interference system or schlieren optics are so high that nonideal conditions result, and the data are spurious. Under these conditions, accuracy may be compromised and recourse to absorption optics with its greater sensitivity is required.

In addition to the monochromator as a variable wavelength light source, the absorption optics system employs a split-beam densitometer, a chart recorder, and makes available a facility for multiple channels or cells. Single sector cells with solvent in one cell and solution in the other, placed in different holes in the rotor allow the scanner to produce directly the optical density of the solution relative to the solvent.

Sedimentation coefficients are readily calculated from both integral and derivative traces. In practice, for biosolutes such as viruses, the position of the maximum ordinate of the gradient curve is a sufficiently accurate measure of the true boundary position. Therefore, most measurements involve the peak of the derivative curve or the 50% level in integral traces. The derivative trace represents the concentration gradient as a function of position in the cell. The integral trace records concentration (optical density) as a function of distance.

5.1.2.2 The Schlieren Optical System

In the 1930s and 1940s the schlieren (streaks, in German) system replaced the rudimentary absorption optics used at that time in sedimentation experiments. This system is based on the interception of light rays that are deflected when passing through an index of refraction gradient. The pattern obtained represents the concentration gradient as a function of radial distance (Figure 5.9). A double sector cell

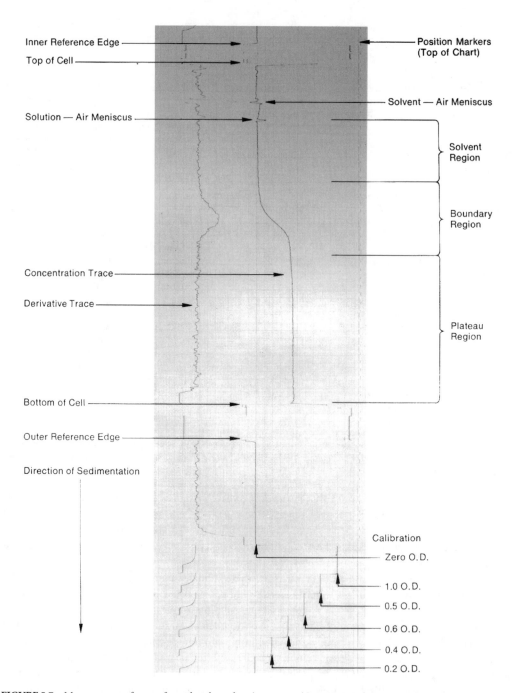

FIGURE 5.7   Measurement of traces from the photoelectric scanner. (Courtesy: Beckman Instruments, Palo Alto, CA.)

permits the baseline to be obtained simultaneously with the patterns (Figure 5.10). The boundary region is affected by diffusion. The lower the diffusion coefficient, the sharper the peak. Viruses, with their relatively large sizes and low diffusion coefficients, give sharp peaks, and can be studied at higher dilution than proteins. With the Model E analytical ultracentrifuge the schlieren optical system is routinely used with all types of sedimentation runs except the moving zone method, with its requirement for exceedingly dilute solutions.

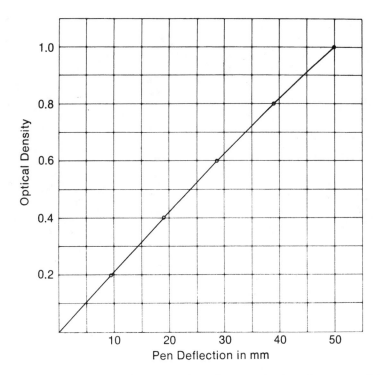

FIGURE 5.8 Typical conversion graph for scanner charts. (Courtesy: Beckman Instruments, Palo Alto, CA.)

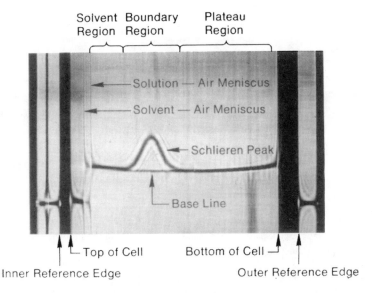

FIGURE 5.9 A typical schlieren pattern showing reference parameters. Refer to text. (Courtesy: Beckman Instruments, Palo Alto, CA.)

5.1.2.3 The Rayleigh Interference Optical System

In the 1950s, because of the demand for enhanced accuracy in sedimentation studies, the Rayleigh interference optical system became a popular means for obtaining data in sedimentation studies. However, the system is more complicated to use than Schlieren optics. It employs double sector cells,

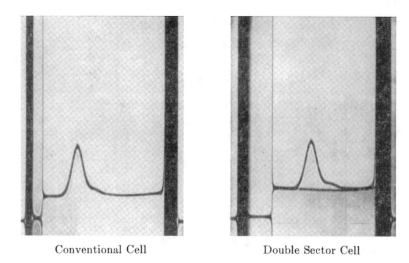

Conventional Cell   Double Sector Cell

FIGURE 5.10   Schlieren patterns obtained with a double sector cell (right) and a conventional cell. The presence of the base line in the experiment using a double sector cell improves the precision in the determination of the faster component. (From Schachman, H. K., *Ultracentrifugation in Biochemistry*, Academic Press, New York, 1959. With permission.)

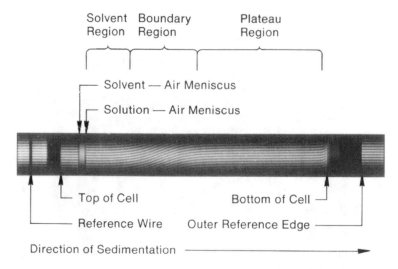

FIGURE 5.11   A typical interference fringe pattern showing reference and solution-solvent relationships. (Courtesy: Beckman Instruments, Palo Alto, CA.)

special optical masks, and critical alignment. The optical pattern is in the form of fringes (Figure 5.11), obtained through the differential retardation of light passing through a solvent and a solution at the same distance from the center of rotation. The fringes represent concentration as a function of radial distance. Rayleigh interference optics cannot be used in isodensity equilibrium or moving zone methods. In addition, the system cannot be used at high concentration or high speeds, because the fringes cannot be resolved. The interference system gives the concentration difference with respect to radius, and in this feature is complementary to the schlieren system, which gives the concentration derivative with respect to radius.

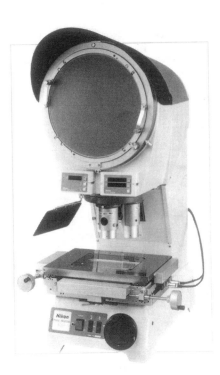

FIGURE 5.12   Optical micrometer. (Courtesy: Nikon Inc., Melville, New York.)

5.1.3   Analyzing the Optical Record[234]

Measurements on the recorded images of optical patterns in the ultracentrifuge require refinement in order to obtain accuracy in the final result. The optical record may be analyzed manually, or by automated procedures, and by online connection where the optical signal is fed directly into a computer.

Manual methods are tedious and time consuming procedures associated with processing and measuring photographic images or traces. However, the details, as presented here, will acquaint the reader with the basics involved in the measuring process.

The usual way to make manual measurements on Schlieren and interferometric records is with an optical microcomparator (Figure 5.12) or a good quality photographic enlarger. A film densitometer, e.g., the analytrol, is used for absorption records (Figure 5.13).

5.1.3.1   Manual Techniques for Schlieren Patterns

Figure 5.9 is a typical schlieren pattern obtained with a double sector cell and interference counterbalance. The portion of the pattern corresponding to the cell is a plot of dn/dr vs. r, where n is the refractive index and r is the radial distance from the center of rotation in centimeters. Changes in refractive index, dn, can be considered to be directly proportional to changes in solute concentration, so that the pattern is effectively a plot of dc/dr vs. r — a plot of the rate of change of concentration with radius.

The schlieren peak represents the boundary region and the maximum ordinate of the schlieren peak is the boundary position. The area circumscribed by the schlieren peak and the baseline is proportional to the total concentration of the solute in the cell. The magnitude of dc/dr is a function of the angle of the schlieren diaphragm (phase plate) as well as the concentration. To compare patterns recorded at different plate angles, the values of dc/dr must be multiplied by the tangent of the phase plate angle used in each case (Table 5.2).

Because of the magnification of the camera lens, all dimensions on photographic plates are larger than the actual dimensions in the cell and counterbalance. The ratio of image cell size is the magnification

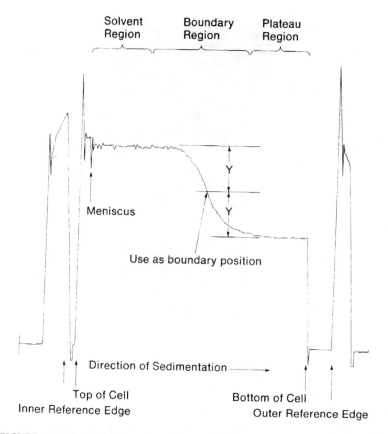

FIGURE 5.13   An Analytrol tracing. (Courtesy: Beckman Instruments, Palo Alto, CA.)

TABLE 5.2   Tangents of Angles

Angle, θ	Tangent θ
30	0.57735
35	0.70021
40	0.83910
45	1.0000
50	1.1918
55	1.4281
60	1.7321
65	2.1445
70	2.7475
75	3.7321
80	5.6713
85	11.430

From Chervenka, C.H., *A Manual of Methods for the Analytical Ultracentrifuge*, Beckman Instruments, Palo Alto, CA, 1969. With permission.

factor, F. This factor must be determined accurately for each optical system of an ultracentrifuge. For computational purposes, in general, all radial dimensions measured on photographic plates are divided by the magnification factor, and the actual dimensions on the scale of the cell are used.

The magnification factor can be determined from photographic images of the calibration cell, from the camera lens magnification photo, or from an experimental photo in which the schlieren counterbalance was used. With the calibration cell photo, the horizontal distance in centimeters between the images of the two vertical lines is measured carefully in the comparator and divided by the actual spacing between the lines, the value for which is stamped on the cell housing. With a schlieren photo, the distance between the images of the two reference edges of the counterbalance is measured and divided by the actual distance. For Model E cells, this distance is usually 1.60 cm. However, the spacing should be accurately measured.

For runs using single sector cells, a baseline must be provided if dc/dr is to be measured. This can be done by rinsing and refilling the cell with solvent, without disassembly, at the end of the sample run, then rerun at the same speed, taking photographs at the same phase plate angle as the sample run. The schlieren pattern obtained is then superimposed, using comparator measurements, upon the sample pattern.

Measurements of radial distances are made relative to the image of the inner reference edge of the schlieren counterbalance. In the case of the Model E, this edge is usually 5.70 cm from the center of the rotor at low rotor speed.

The plate is placed on the comparator carriage with the emulsion side up and oriented so that the image, as seen in the comparator view screen, is right side up and the imaginary center of rotation is to the left as one faces the comparator. The top of the cell is now the left side of the image. The plate is rotated so that a meniscus image is exactly parallel to the vertical cross hair of the view screen as the plate is moved up and down across the view screen.

Trautman[235] proposed a procedure to measure radial distance. The plate is positioned laterally, relative to the horizontal scale of the comparator, so that an imaginary zero of the scale is at the imaginary center of rotation of the cell image on the plate. The comparator scale then reads directly in radial distance, with dimensions F times the actual centrifuge dimensions. Thus, if the magnification factor is 2.00, the plate is set so that the inside reference edge is 11.40 cm (2.00 times 5.70, the actual distance between the reference edge and the center of the rotor) from the imaginary zero of the comparator scale. If the comparator scale is short, the imaginary zero could be set 10 cm from the actual zero of the scale. The plate is moved along the carriage until the reference edge is aligned with the vertical cross hair when the carriage is at 1.40 cm on the comparator scale— 11.40 cm from the imaginary center. To obtain the radial distance r, to any position in the cell, such as the meniscus, the horizontal carriage is moved until the vertical cross hair is aligned with the image of the meniscus and 10.00 cm is added to the comparator scale reading. This value is then divided by the magnification factor. All comparator distances are read to the nearest 10 μm.

5.1.3.2 Comparator Readings for Interference Patterns

Figure 5.11 is a typical reference fringe pattern obtained using a double sector interference cell and counterbalance. Any fringe in the portion of the pattern corresponding to the cell is a plot of concentration, c, vs. r, again assuming that concentration is a linear function of refractive index. The vertical shift of fringes between one radial position in the cell image and another is a direct measure of the change in concentrations of solute between the two positions in the sample sector of the cell, assuming uniform refractive index in the reference sector. Since the fringes are uniformly spaced vertically, the fringe shift usually is determined by a horizontal count, which is described below. These values of fringe shift, Δj, are values of concentration with units of fringes, and the units are used for computations. The total fringe shift across a boundary, J, is a measure of the total concentration change across the boundary, and gives the concentration in the plateau region. In general, for proteins, a shift of one fringe corresponds to a concentration of approximately 0.25 mg/ml in a 12 mm cell. To locate the boundary position in interference fringe patterns, one uses the half-concentration point. This point is the radial position of the midpoint of the shifted fringes.

The plate is placed on the comparator carriage with the fringes in the reference hole aligned with the horizontal axis. The plate is rotated so that after the horizontal cross hair is centered on a fringe in one

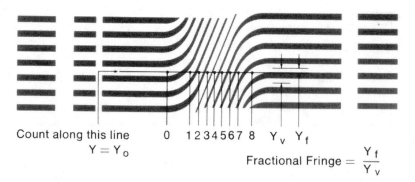

Count along this line   0   1 2 3 4 5 6 7 8 Y_v Y_f
$Y = Y_0$

$$\text{Fractional Fringe} = \frac{Y_f}{Y_v}$$

FIGURE 5.14   Determination of the fractional fringe. (Courtesy: Beckman Instruments, Palo Alto, CA.)

reference hole image, the carriage can be moved laterally and the cross hair will still be centered on the corresponding fringe in the other reference hole image. It should be noted that the number of fringes in the two reference hole images is different. The magnification factor is checked as described for schlieren measurements, and for the Model E, the value of 1.67 cm is taken as the distance between the center of the reference wire image and outside reference edge of the interference counterbalance.

Radial distances are determined as with schlieren plates, except that 5.62 cm is used as the nominal distance between the center of the reference wire image and the center of the rotor. Radial distances on the plate are measured to the nearest 10 μm. For the determination of fractional fringe numbers, distances are estimated to 1 μm on the comparator scale.

5.1.3.3 Determination of Vertical Spacing Between Fringes

The vertical fringe spacing is an important constant that is determined for each centrifuge. To determine this value, the vertical distance between the center of one bright fringe and the next one at a position in the cell image of an interference plate is measured where the fringes are straight. To increase the accuracy of the determination, one measures the distance over as many fringes as are clear in the image, and then divides by the number measured over. The measurement is repeated several times and on several plates and the results are averaged. The vertical fringe spacing should be known accurately within 5 μm. Measurements of fringe spacing are not corrected for camera lens magnification.

With the plate aligned as described above, the carriage is moved laterally until the vertical cross hair is at the radial position desired. For sedimentation equilibrium plates this will be the meniscus. For boundary runs, such as the synthetic boundary determination of initial concentration, the starting position will be in the solvent region of the cell where the fringes are horizontal, but near to the boundary. In either case, the carriage is moved vertically until the horizontal cross hair is in the center of a clear fringe at the intersection of the cross hairs. The vertical position of the carriage is recorded as y_0 (Figure 5.14). For the boundary run, the carriage is moved laterally, counting each bright fringe that is passed by the intersection of the cross hairs, until the plateau region of the image is reached. For an equilibrium run, the counting continues to the bottom of the liquid column. In each case, the fraction of a fringe remaining is determined by moving the carriage vertically until the intersection of the cross hairs is again centered on the last whole fringe counted. The vertical scale reading is read and recorded. The distance in centimeters that the carriage was moved vertically from y_0 is divided by the vertical fringe spacing, again in centimeters. This value was determined previously as described above. The result is the fractional fringe, which is added to the whole fringe count to give the total fringe shift.

In reading sedimentation equilibrium plates there is a problem in locating fringe positions at the top and bottom of the fluid column. This requires an extrapolation to locate the position of a fringe in the center of the meniscus image. The fringes at the top of the cell are generally not too steep where they intersect the meniscus image, so the extrapolation is made without too much difficulty. At the bottom of the cell, however, even in the case where oil is used to raise the sample fluid column off the bottom of the centerpiece, fringe positions are much more difficult to locate. The fringes do not extend in the meniscus image, so that there is a wide region in which the position of the fringes is uncertain. In addition, the fringes are generally quite steep. It is generally possible to make the extrapolation by eye and obtain results that are satisfactory for most work. A refined procedure has been published to aid this process.[236] The Model E manual advises the use of transparent material with a straight scribe line be held on the comparator screen to aid the eye in following the extension of the fringe pattern to the nominal bottom meniscus position.

Interference fringes formed at high rotor speeds may slope slightly due to cell distortion. This is usually a small effect, but for accurate work a correction can be made. For sedimentation equilibrium runs, the fringes are photographed immediately after the rotor attains operating speed, and before much redistribution of solute has taken place. For other kinds of experiments, the cell is emptied after the run, rinsed, and refilled with water, without disassembly. It is run at low speed, again rinsed, then refilled exactly as in the original run, except that water is used instead of solvent and solution. The cell is then run up to the same speed used in the original run, and the fringes are photographed. The resulting plate is aligned in the comparator, and a record made of the fringe displacement vs. radial position in the cell. The values obtained are then added or subtracted as required from the values for displacement for the analytical run.

5.1.3.4   Measurements of Absorption Records

Recording an ultracentrifuge run on UV film employs a film densitometer. The patterns obtained are plots of optical density, with decreasing absorbance from bottom to top, vs. radial position. Most applications of the UV optical system require only measurements of radial positions of special features in the image — menisci, boundary centers, or band centers. Measurement of radial positions is analogous to the measurement of schlieren or interference plates, except that a centimeter scale or a count of graph lines is used to determine distances on the chart. The magnification factor, the product of the magnification of the optical system and that due to the Analytrol, is determined by measuring the distance between the reference edge images and dividing by the counterbalance dimension, usually 1.60 cm. Since the trace of a sharp optical edge, such as a reference edge, slopes because of the response time of the recorder, positions of vertical edges are taken to be located near the start of the vertical pen stroke. As described above, the magnification factor determined in this way is multiplied by 5.70 cm to obtain the distance to the imaginary center of rotation. A mark is made on the chart at some convenient interval from this center. Then measurements of features of the trace are referenced to this mark. When the resulting numbers are divided by the magnification number, actual radial distances in the cell are obtained. Boundary positions are located as noted in Figure 5.13. A horizontal line is drawn on the chart halfway between the tracing of the solvent region and the plateau region of the cell, and the point of intersection with the tracing of the boundary is used as the center of the boundary.

5.1.3.5   Measurements on Traces from the Photoelectric Scanner[233]

In the late 1950s, photoelectric scanners were incorporated into the optical system of the analytical ultracentrifuge to record absorption runs. These photoelectric scanners were analog instruments which produced recorder traces of optical density as a function of radial level in the cell. The optical densities were obtained through the use of holding circuits which stored the respective signals after conversion to their logarithmic values. In the 1970s efforts were made to redesign the scanner so that double beam measurements are made directly on the individual light pulses. Requiring an online computer, this modification has the decided advantage of producing data in digital form.

Traces obtained with photoelectric scanner are plots of optical density, with increasing absorbance from bottom to top, vs. radial position (Figure 5.7). Measurement of radial distances is similar to the procedure described for Analytrol curves. For accurate work, the manual recommends measuring distances on the chart relative to the position markers made by the pen at the top of the chart. These markers are generated by the rotation of the lead screw of the photomultiplier carriage, and indicate exactly the movement of the carriage, independent of the scanning speed or the recorder chart speed. These blips are inconvenient to use as a scale, and for most purposes measurements made directly on the chart with a centimeter scale are sufficiently accurate.

The magnification factor is determined as noted, except that an interpretation of the chart pattern at the reference edges is required resulting from the double-beam feature of the scanner.

The trace shows a double stroke of the pen — one down and one up — at each reference edge (Figure 5.7). The correct position of either reference edge is indicated by the stroke of the pen in the negative direction from zero outside diameter (O.D.). This part of the trace slopes due to the finite width of the scanner slit, so that the exact position is indicated when the pen reaches –0.3 O.D. However, with the usual setup the pen reaches its lower limit before this. Therefore, the feature used most conveniently is the position where the pen is at the end (or start) of the negative O.D. stroke at the lower pen limit. Any error resulting from the use of this position instead of a more exact one is negligible, so long as the same position is used consistently and the magnification factor is determined correctly.

The manual notes that the sharp vertical stroke of the pen to the infinite O.D. limit should not be used as the reference edge position. The reason is that the pen stroke is initiated by the stray light discriminator circuit of the scanner, not by the signal circuits. The nominal distance between the reference edges of the scanner counterbalance is 1.62 cm and the nominal distance of the inner reference from the center of rotation is 5.70 cm. The actual distances in the counterbalance and rotor should be measured.

Values for absorbancies in the cell are obtained by relating the pen position in the trace to the calibration steps, formed at the beginning of the trace. If the calibration steps are not uniform in size, a calibration graph of pen deflection of each calibration step, in millimeters, vs. optical density, is made as shown in Figure 5.8. This graph is used to convert pen deflections in radial positions of interest into optical density units. The measurements of pen deflections are made relative to a zero O.D. line drawn across the chart connecting the zero O.D. levels in the two counterbalance images.

5.1.3.6   Off-Line Analysis

Rowe et al. described methods for off-line analysis of sedimentation velocity and sedimentation equilibrium patterns.[237] With Rayleigh interference optics, the classical approach, as noted above, has been to give a pattern in which the displacement of fringes in a direction (z) normal to radial is a linear function of the concentration increment at the radial position in question. The fringes are equally spaced and parallel and a scan across them in the z direction yields a sinusoidal intensity function whose phase is a measure of the noninteger part of the fringe shift.

A simple, fast, and stable algorithm was developed for deriving the phase shift from the intensity function. The latter is logged from the photographic record of the fringe pattern, using a commercial scanning densitometer, the LKB (Bromma, Sweden) 2202 laser densitometer. If Q fringes are contained within the window analyzed, an iterative frame shift is performed within the data set, to maximize the Fourier coefficient of order Q. The method is therefore a null method, which searches for the frame shift that will set the phase term to zero. This algorithm yields estimates for the fringe increment whose precision is not a function of the latter parameter itself. The off-line algorithm, ANALYSE2, was written in Turbo Pascal from c vs. r records. It permits the analysis of data from a two-dimensional data acquisition system as opposed to a series of individual one-dimension scans. Sophisticated search procedures were incorporated to ensure that the system reproducibly and stably finds the correct fringe intensity maximum for a full two-dimensional record.

5.1.3.7   Online Analysis

Automated photographic plate readers still require that photographs be taken, processed, and aligned before data analysis can be performed. The Beckman Optima XLA Analytical Ultracentrifuge incorporates as standard procedure the logging of absorbance data to disk, formatted as ASCII files. The same facility will apply to Rayleigh interferometric and to schlieren patterns as and when the modern analytical ultracentrifuge incorporates the appropriate optical systems.[237]

In the U.S. a method for online data acquisition of the Rayleigh inteference optical system has been reported by a group of investigators representing the National Ultracentrifugation Facility at the University of Connecticut, Storrs, the Boston Biomedical Research Institute, and the University of New Hampshire.[238] A Rayleigh interference image is well suited for television camera-based data acquisition. Two types of automated Rayleigh interferometers were adapted to the Beckman Model E analytical ultracentrifuge. One type of system relays and magnifies the Rayleigh interference image from the usual photographic plane to a television camera located behind this plane. The other system uses a redesigned camera–cylinder lens combination to create a radially demagnified Rayleigh interference image of the cell on the television camera sensor located on the original cylinder lens mount. Black and white solid-state television cameras are well suited to the acquisition of interference images by virtue of the close and precise spacing of the light-sensitive cells (pixels), adequate light sensitivity, insensitivity to blooming or damage from bursts of light, and their generally rugged nature.

Each system uses a laser-diode light source and a commercially available computer interface to acquire the interference image. The image is aligned so that the columns of pixels on the camera sensor are perpendicular to the radial direction and, therefore, each column corresponds to a radial position. A single-frequency, discrete Fourier analysis at each column of pixels provides the fractional fringe displacement in a manner somewhat similar to that first described by DeRosier and coworkers.[239] The filtering performed by the Fourier analysis minimizes the need for accurate reporting of the fringe intensities and reduces the requirement for high-quality images. Keeping track of the integral number of fringes traversed enables the image to be converted to fringe displacement vs. radial position.

It was reported that both optical systems provide excellent performance. Image acquisition requires 1/30th of a second and image processing less than 10 seconds. The precision of fringe displacement measurements is comparable to that of automated plate readers.

Data translation computer interfaces for the MicroVax II (model DT2651), IBM-PC (model DT2851), and Apple MacIntosh II (model DT 2255) have been used to digitize the video image. Software was written in FORTRAN or C and linked to the data translation DT-IRIS subroutine library. These programs are available from the authors.

5.1.4 Determination of Concentration and Spectrophotometric Parameters

In sedimentation experiments a knowledge of the initial concentration of solute is frequently required. Conveniently, these concentrations should be known in the same units as those used for a particular kind of run: optical density for absorption optics, the number of fringes for interference optics, and square centimeters for schlieren optics. For absorption optics initial concentrations are obtained from film or scanner records made early in the run, and no special determination is required.

In boundary experiments, the area under the schlieren peak or the total number of fringes shifted is proportional to the concentration change across the boundary. Thus, in sedimentation velocity runs, the concentration of a solution can be determined by measuring the peak, or by counting the fringes shifted through the boundary. The use of sector shaped cells results in a decrease in concentration in the plateau region of the cell during a run. For a sedimenting schlieren peak the area will be found experimentally to decrease as it moves through the cell. The plateau concentration at any time, c_t, can be compared to the initial concentration, c_o, by the equation

$$c_t = c_o \left(\frac{r_o}{r_t} \right)^2 \tag{5.1}$$

The position of the meniscus gives the radial distance to the boundary at zero time, r_o, while r_t is measured as the radial distance to the maximum ordinate of the schlieren peak at time t.

The use of a synthetic boundary cell minimizes the effect of radial dilution and runs of only a few minutes duration are required. In practice the double sector capillary-type synthetic boundary cell is commonly employed for the determination of the initial concentration. This cell can be used for both schlieren and interference optics. In the left-hand sector is placed solvent representing the dialyzate of the sample solution. The sample solution is placed in the right-hand sector. Upon acceleration of the rotor the boundary formed is very sharp and the schlieren pattern or interference fringes on either side

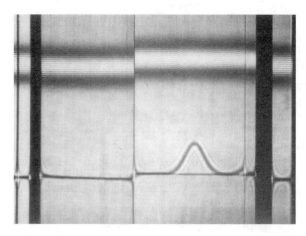

FIGURE 5.15   Combined schlieren (bottom) and interference (top) optics photo of a synthetic boundary experiment. (Courtesy: Beckman Instruments, Palo Alto, CA.)

of the boundary should be straight and horizontal. The spreading of the boundary, if it occurs, should be very slow and symmetrical — with fringes quite clear, or the schlieren pattern should approximate an equilateral triangle (Figure 5.15).

For each plate exposure, the radial distance to the meniscus, the radial position of the boundary, and the area under the peak for schlieren optics (discussed below) are determined, or the fringe shift for interference optics. Using the boundary position of the initial photo as r_o, each value of area or fringe shift is corrected for radial dilution, by multiplying each value by $(r_t/r_o)^2$. For interference optics, the Model E manual suggests that a simpler alternative is to make a plot of measured fringe shift for each photo vs. time and to extrapolate the plot to zero time. If there is no measurable boundary movement, the run can stand on its own.

The Beckman Optima analytical ultracentrifuges possess increased sensitivity and wide wavelength range in terms of absorbance optics. The absolute concentration is available in principle at any point. Samples may be examined in concentrations too dilute for schlieren or interference optics. With proteins, measurements below 230 nm allow examination of concentrations 20 times more dilute than can be studied with interference optics. Concentrations as low as several micrograms per milliliter are now accessible.[227]

The Beckman Optima XL-I model is equipped with both Rayleigh interference and UV absorbance optical systems. The extinction coefficient can be measured readily. Layering solvent over a dilute solution of protein in a synthetic boundary cell permits the measurement of the absorbance and the number of fringes. The latter quantity yields with high accuracy the concentration in milligrams per milliliter and the absorbance can then be converted into the extinction coefficient on a weight/volume basis.[240]

In another study the refractive increment of a protein was used to determine its concentration. The absorption spectra and interference pattern were obtained with the XL-I model using a synthetic boundary cell. The absorbance and fringe displacement of the protein were determined by subtracting the mean value of the baseline from the plateau. The specific refractive increment value was obtained from the literature. The measured fringe displacement was converted into a concentration value. The determined concentration together with the measured absorbance at 280 nm yielded the absorptivity. From the known molecular weight the molar extinction coefficient was calculated.[241]

5.1.5   Measurement of Areas of Schlieren Peaks[234]

Figure 5.16 illustrates a rectangular approximation method for measuring areas under schlieren peaks — graphical integration. The approximate area of any trapezoidal portion of the pattern, abcd, is equal to the height of the pattern measured at position 4, times the increment of radial distance, cd. The area of

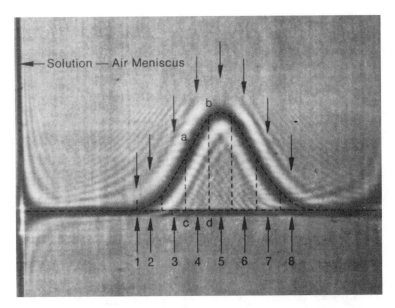

FIGURE 5.16   Rectangular approximation method for areas of schlieren patterns. (Courtesy: Beckman Instruments, Palo Alto, CA.)

the entire pattern is then closely approximated by the sum of all the trapezoidal portions. In the comparator the plate is aligned so that a meniscus image is parallel to the vertical axis, and the vertical crosshair is at the position where the center of the schlieren pattern meets the center of the baseline (position 1). A convenient increment of radial distance is to use 0.1 cm on the comparator scale. The horizontal carriage of the comparator is moved 0.05 cm (position 2) and the vertical distance from the baseline to the schlieren pattern dc/dr, is measured in centimeters. The values for dc/dr and the radial distance are recorded. The carriage is then moved 0.1 cm (position 3) and the measure of dc/dr is again recorded. This operation is repeated until the entire pattern is traversed. The area under the peak is now equal to the sum of all the values of dc/dr multiplied by 0.1/F, to correct radial distances to actual cell dimensions. The value obtained for the area, and thus the concentration, will have the units of cm^2. For all usual computations using schlieren optics, the dimension of cm^2 for concentration can be retained and such values can be substituted directly into equations.

5.1.6   Interconversion of Schlieren and Interference Fringe Concentrations

A variation of the synthetic boundary procedure is to make all determinations using interference optics, even when concentrations are required for a companion experiment with schlieren optics. By substituting the simple count of interference fringe shift across the boundary, the routine integration of schlieren patterns can be avoided. The measured fringe shifts can then be converted into equivalent schlieren areas by the use of a conversion factor (CF). The factor is a constant for each ultracentrifuge and is independent of the nature of the solute. It may be determined as follows. A synthetic boundary run is made using a convenient solute such as sucrose in a concentration between 0.5 and 1.0%. The pattern is recorded using both schlieren and interference optics (Figure 5.15). The values for schlieren area (A) and fringe count (J) obtained from the run are substituted into the following equation:

$$CF = \frac{A\left(\text{schlieren}\right) \times \tan \theta}{J} \tag{5.2}$$

Replicate determinations are made and the results are averaged.

An alternative to the above procedure involving the synthetic boundary cell is the use of an accurate differential refractometer, to determine the required initial concentration of a sample solution. A knowledge of the optical constants of the ultracentrifuge is required: the optical lever arm and the magnification of the camera and cylindrical lens. Such values are supplied by the manufacturer or can be obtained by use of the calibration cell that comes with the instrument. Directions are obtained in the manual.

5.1.7 Sedimentation Velocity

Experiments of the sedimentation velocity type make up the most common usage of the ultracentrifuge. In a run, with a few milligrams of virus and a relatively short period of time, important information is obtained: the presence of macromolecular species is observed, the amount of each species can be estimated, sedimentation coefficients can be calculated, molecular size may be estimated, the extent of molecular heterogeneity is revealed, and the relative amounts of components can be estimated.

5.1.7.1 Sedimentation Theory[242,243]

In sedimentation, a force, F, acting on a mass m causes an acceleration, a, so that F = ma. The acceleration in a centrifugal field is $\omega^2 r$, where ω is the angular velocity and r is radial distance. The force equation becomes $F = m\omega^2 r$.

The force moves a particle through a viscous medium but encounters a resistant frictional force, F′, pushing the particle back. F′ is directionally proportional to the velocity, v, of the particle and f, the translational frictional coefficient, a function of the size and shape of the particle and the viscosity of the solution. Since F′ works in a direction opposite to F, F′ = –fv. In a centrifuge run, with acceleration of the particle, a point in time is reached where the accelerating force is equal in magnitude, though opposite in direction, to the frictional force. At this point, no net force on the particle occurs. It will no longer speed up, nor slow down, but continues to move through the solvent at a well-defined velocity.

$$F + F' = 0 \tag{5.3}$$

$$m\omega^2 r - fv = 0 \tag{5.4}$$

Rearranging Equation 5.4 yields

$$\frac{v}{\omega^2 r} = \frac{m}{f} \tag{5.5}$$

Equation 5.5 does not take into consideration the buoyancy of the medium. The quantity m must be related to the buoyant mass, mb, i.e., the mass of the particle minus the mass of an equal volume of the liquid medium. The volume of the particle is equal to $m\bar{v}$, where $\bar{v}$ is the partial specific volume of the hydrodynamic particle, i.e., the particle plus any bound substance (e.g., water, salt) traveling with it under the action of the centrifugal field. The quantity $\bar{v}$ is in ml/g. If the volume $m\bar{v}$ is multiplied by the density of the medium, ρ, which has units, g of solution/ml, the product $m\bar{v}\rho$ will be the weight of an equal volume of solution. The mass of the hydrodynamic particle minus the mass of an equal volume of solution becomes

$$mb = m - m\bar{v}\rho = m(1 - \bar{v}\rho)$$

Thus, Equation 5.5 becomes

$$\frac{v}{\omega^2 r} = \frac{mb}{f} = \frac{m\left(1 - \bar{v}\rho\right)}{f} \tag{5.6}$$

In centrifugation experiments, the molecular weight, M, is calculated more commonly than the mass, m. M and m are related through Avogadro's number N, by $m = M/N$ and Equation 5.6 becomes

$$\frac{v}{\omega^2 r} = \frac{M(1 - \bar{v}\rho)}{Nf} \tag{5.7}$$

Each side of Equation 5.7 for a given particle in a particular medium will be constant: the left side of the equation is the velocity per unit centrifugal field, and the right side of the equation reflects the size, shape, and molecular weight of the particle. The quantity $v/\omega^2 r$ may be obtained from the centrifuge run, and is referred to as the sedimentation coefficient, s:

$$s = \frac{v}{\omega^2 r} = \frac{M(1 - \bar{v}\rho)}{Nf} \tag{5.8}$$

From Equation 5.8, the velocity, v, may be written in the calculus equivalent, dr/dt, and

$$s = \frac{1}{\omega^2 r} \frac{dr}{dt} = \frac{1}{\omega^2} \frac{d\ln r}{dt} \tag{5.9}$$

To obtain a parameter independent of the rotor or speed used, the velocity per unit field strength is calculated

$$s = \left(\frac{10^{13}}{\omega^2}\right) \frac{d\ln r}{dt} \tag{5.10}$$

where the factor 10^{13} converts the result to Svedberg units, and $\omega = 2\pi \, (\text{rpm})/60$ where (rpm) is the number of rotor revolutions per minute.

Generally, the sedimentation value measured is converted to standard conditions, i.e., a value of $s_{20,w}$, which is the sedimentation coefficient obtained in a medium having the same equation density (ρ) and viscosity (η) as water at 20°C. The equation for the conversion is[244]

$$s_{20,w} = s_{obs}\left[\left(1 - \bar{v}_b\rho_{20,w}\right)/\left(1 - \bar{v}_b\rho_T\right)\right]\left(\eta_r/\eta_{20}\right)\left(\eta/\eta_0\right) \tag{5.11}$$

where ρ_T is the solvent density and $\rho_{20,w}$ that of water at 20°C, η_r/η_{20} is the relative viscosity of water at the temperature T with respect to 20°C, and η/η_0 is the relative viscosity of the solvent with respect to water.

From Equation 5.8, the first fundamental equation for the calculation of molecular weight from sedimentation data may be derived. However, the concept of diffusion must be taken into consideration and involves a determination of the diffusion coefficient, D.

It is generally accepted for infinitely dilute solutions that the frictional coefficient, f, in sedimentation is the same as that present in transport by diffusion.

The diffusion coefficient is related to the frictional coefficient, f, by

$$D = RT/Nf \tag{5.12}$$

Substituting in Equation 5.8, the fundamental equation for M in sedimentation velocity, the Svedberg equation, is obtained:

$$M = RTs/D(1 - \bar{v}\rho) \tag{5.13}$$

TABLE 5.3   Typical Data for Calculation
of Sedimentation Coefficient

Photo No.	Time (min)	x Scale Reading (cm)	$\log_{10} x$
	0	12.380 (meniscus)	1.0927
1-1	8.3	12.537	1.0982
1-2	16.3	12.686	1.1033
1-3	24.3	12.840	1.1086
1-4	32.3	12.993	1.1137
1-5	40.3	13.140	1.1186
2-1	48.3	13.297	1.1238
2-2	56.3	13.456	1.1289
2-3	64.3	13.611	1.1339
2-4	72.3	13.766	1.1388
2-5	80.3	13.923	1.1437

From Chervenka, C.H., *A Manual of Methods for the Analytical Ultracentrifuge,* Beckman Instruments, Palo Alto, CA, 1969. With permission.

R is the universal gas constant, 8.313×10^7 ergs per degree per mole, and T is the absolute temperature.

The Svedberg equation applies only at infinite dilution, and corrections are required, as for equations for sedimentation equilibrium, for nonideal solutions at finite concentrations. Moreover, the Svedberg equation is restricted to two-component systems containing a solute and a solvent.[245]

5.1.7.2   Determination of Sedimentation Coefficients

The rate of sedimentation of solute molecules subject to a high centrifugal force yields the sedimentation rate of the average of all macromolecules. The process involves measuring the recorded position of the boundary in the centrifuge cell as a function of time. In absorption optics, the boundary is a recording of the optical density from the photoelectric scanner, and in schlieren optics, the peak. (The use of interference optics is not particularly useful for the determination of sedimentation rates.)

For schlieren plates one measures the radial distance of the sample meniscus and of the boundary from the center of rotation, for each recording taken during the run (Figure 5.9). A comparator is generally used to obtain scale readings in centimeters. Then a table is made, as shown in Table 5.3. Each scale reading, r, designated x in the table, corresponds to an elapsed time in minutes.

The sedimentation coefficient is conveniently determined by plotting $\log_{10} r$, the boundary measured in centimeters, as it moves from the meniscus of the solution to the bottom of the cell, vs. the time in minutes (Figure 5.17). The slope is obtained and related to other factors in the sedimentation equation.

From the slope in Figure 5.17 a value of 6.33×10^{-4} is obtained. The slope value is multiplied by the constant for the run to obtain the sedimentation coefficient. From Table 5.4, $2.303/60\omega^2$, at 52,000 RPM (speed assumed), is 1.295×10^{-9}.

From Equation 5.9, $s = (1.295 \times 10^{-9}) \times (6.33 \times 10^{-4}) = 8.20 \times 10^{13}$. As noted above, a Svedberg unit is equal to 1×10^{-13}, and the sedimentation coefficient equals 8.20 S.

To negate any dependence of the sedimentation rate on concentration, sedimentation values are plotted vs. a series of concentrations of the biosolute, and extrapolated to zero concentration-infinite dilution. The extrapolation to zero concentration should represent as short an interval as possible (Figure 5.18).

For ultraviolet absorption studies, to determine the sedimentation coefficient the photoelectric scanner is particularly useful.[233] For systems requiring measurements at very high dilution it is unrivaled. Most determinations of sedimentation coefficients involve measurements of the rate of movement of boundaries, and with the photoelectric scanner, both the integral trace representing concentration (actually optical density) as a function of distance and the derivative trace corresponding to the concentration gradient as a function of position in the cell are obtained.

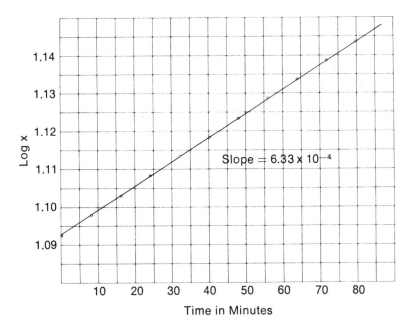

FIGURE 5.17 Plot of log x vs. t for calculation of sedimentation coefficient. (Courtesy: Beckman Instruments, Palo Alto, CA.)

Sedimentation coefficients are calculated from both integral and derivative traces. The position of the maximum ordinate of the gradient curve is an accurate measure of the true boundary position. Therefore, most measurements involve the peaks of the derivative curve or the 50% level in the integral traces. As for schlieren patterns, a plot is obtained of the boundary position as a function of time.

As noted above, for sedimentation velocity experiments it is essential that the cells holding the sample and solvent be sector-shaped, with walls aligned along the radii of the rotor. This prevents sedimenting particles from colliding with the walls (wall effects).

Sedimentation velocity experiments are normally of a type called boundary sedimentation.[244] The experiment begins with a sample mixed uniformly throughout the cell, so that a plot of concentration vs. radius is a horizontal line (C(r) = constant). As sedimentation proceeds, molecules are depleted from the top of the solution column. This results in the formation of a trailing boundary for the concentration distribution.

The ability to resolve boundaries is proportional to $\omega^2 rL/\theta$ where L is the column length, and θ is the width of the section. Thus, a long, narrow solution column is generally preferred for highest resolution. Typical solution columns for sedimentation velocity runs hold 0.45 ml.

Figure 5.13 shows typical data acquired during a boundary sedimentation experiment. The data reflect a plot of the solute concentration as a function of radial distance, or C(r). A number of key features of the data are pointed out in the figure. The pair of sharp peaks indicate the positions of the menisci — sample and solvent. In the boundary region the solute concentration increases rapidly to a reasonably constant value in the plateau region. Most of the information in a sedimentation velocity experiment is taken from analysis of the boundary. In a simple sedimentation run involving one component, the boundary will be sharp, and the sedimentation coefficient can be derived from the motion of the boundary midpoint r_b. This is most readily calculated from the slope of the equation, ln $(r_b) = (\omega^2 s)$ t, where the time, t, is plotted in seconds.

In more complex analyses involving two or more components, the boundary will be divided into two or more rising segments. Assuming each component has the same extinction coefficient, the relative heights of the boundary segments represent the relative concentration of each component. The radial

TABLE 5.4 ω^2 and $2.303/60\omega^2$ Data

Speed Selector Setting — LO Range	ω^2	$\dfrac{2.303}{60\omega^2}$	Speed Selector Setting — HI Range	ω^2	$\dfrac{2.303}{60\omega^2}$
800	7.016×10^3	5.471×10^{-6}	8,000	7.016×10^5	5.471×10^{-8}
900	8.879	4.323	9,000	8.879	4.323
1,000	1.097×10^4	3.499	10,000	1.097×10^6	3.499
1,100	1.327	2.892	11,000	1.327	2.892
1,200	1.579	2.431	12,000	1.579	2.431
1,300	1.853	2.071	13,000	1.853	2.071
1,400	2.149	1.786	14,000	2.149	1.786
1,500	2.467	1.556	15,000	2.467	1.556
1,600	2.807	1.367	16,000	2.807	1.367
1,700	3.168	1.212	17,000	3.168	1.212
1,800	3.553	1.080	18,000	3.553	1.080
2,000	4.387	8.749×10^{-7}	20,000	4.387	8.749×10^{-9}
2,200	5.308	7.231	22,000	5.308	7.231
2,400	6.317	6.076	24,000	6.317	6.076
2,600	7.411	5.179	26,000	7.411	5.179
2,800	8.598	4.464	28,000	8.598	4.464
3,000	9.870	3.889	30,000	9.870	3.889
3,200	1.123×10^5	3.418	32,000	1.123×10^7	3.418
3,400	1.268	3.027	34,000	1.268	3.027
3,600	1.421	2.701	36,000	1.421	2.701
4,000	1.755	2.187	40,000	1.755	2.187
4,400	2.123	1.808	44,000	2.123	1.808
4,800	2.527	1.519	48,000	2.527	1.519
5,200	2.965	1.295	52,000	2.965	1.295
5,600	3.439	1.116	56,000	3.439	1.116
6,000	3.948	9.722×10^{-8}	60,000	3.948	9.722×10^{-10}
6,400	4.49	8.545	64,000	4.492	8.545
6,800	5.071	7.569	68,000	5.071	7.569
7,200	5.685	6.752	72,000	5.685	6.752

From Chervenka, C.H., *A Manual of Methods for the Analytical Ultracentrifuge*, Beckman Instruments, Palo Alto, CA, 1969. With permission.

motion of each segment can be analyzed independently to determine each sedimentation coefficient. In the case of a very broad boundary a heterogeneous sample is indicated.

As an alternative representation, the data may be presented as the derivative of the concentration function, or dc/dr. In this consideration, each boundary segment appears as a discrete peak, and the sedimentation coefficient is obtained from the radial motion of these peaks. The relative concentration of each sample component is determined from the area under each peak.

In the plateau region, particles at greater radii will move faster than those closer to the center of rotation, thus pulling away from the latter.[244] Moreover, as the experiment progresses, particles beginning near the outermost portion of the solution column will be pelleted against the outer wall of the sample cell, and will be replaced by particles from nearer to the center of rotation. These latter particles enter a progressively increasing volume as they migrate outward through the sector-shaped cavity, and thus become more dilute. This phenomenon of radial dilution accounts for the gradual decrease in concentration in the plateau.[246]

5.1.7.2.1   Concentration Dependence.   The sedimentation coefficient should be obtained over a range of solute concentrations and extrapolated to infinite dilution. Highly asymmetric molecules, or molecules forming associating systems, will show concentration dependence of s.[247]

5.1.7.2.2   Molecular Volume and Extension.   Highly asymmetric molecules tend to occupy a disproportionately large volume stemming from their rotational motion in solution. The net effect is to prevent

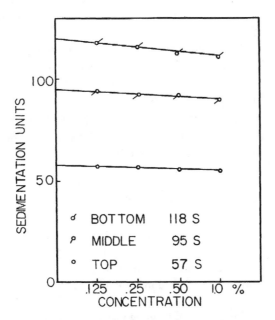

FIGURE 5.18   Sedimentation coefficients for the components of squash mosaic virus. (Data from Reference 257.)

solvent molecules from approaching them, increasing the apparent viscosity of the solvent and reducing the sedimentation rate of the asymmetric solute. Thus, the observed sedimentation coefficient for a highly asymmetric molecule can decrease markedly with increasing concentration.

If macrosolute molecules tend to dimerize or form higher-order associating systems, the sedimentation coefficient may increase with solute concentration. These associating systems are best studied by methods of sedimentation equilibrium.

5.1.7.2.3   The Johnston-Ogsten Effect for Mixtures.   In 1946, Johnston and Ogston observed that at high concentrations, fast-moving macrosolute molecules must move through a layer of slow-moving macrosolutes as well as solvent. The slow-moving species increase the apparent viscosity of the solvent, leading to a concentration-dependent decrease in the sedimentation coefficient. This effect is particularly evident for asymmetric molecules.[248]

5.1.7.2.4   Speed Dependence.   The s value may be affected by speed dependence. The observed sedimentation coefficient for some solutes may increase with increasing rotor speed. Speed dependence is sometimes observed when sedimenting very large, asymmetric molecules, or highly polymerized, but dissociable, molecules. This phenomenon is believed to result from a wake left behind macrosolutes moving through the solution column, clearing buffer and salt molecules from the medium. This permits an increased velocity for trailing macrosolute molecules, resulting in the formation of macromolecular aggregates. Where speed dependence is observed, it can be avoided by working at the lowest practical rotational speeds.

5.1.7.2.5   Solvent Effects.   When charged macrosolutes, such as proteins or nucleic acids, are centrifuged through a polar solvent, they move more rapidly than the solvent counterions that normally envelope them. This results in a charge separation and potential difference that slows the macrosolute molecules, and results in a decrease in the observed sedimentation coefficient. This effect is commonly avoided by using ionic strength in excess of 50 mM.

5.1.7.2.6   Viscosity and Density.   Very dense or viscous solvents will reduce the observed sedimentation coefficient by increasing the forces of buoyancy and frictional drag experienced by the macrosolute. Sedimentation coefficients are therefore conventionally expressed in terms of a standard solvent, water at 20°C. An observed sedimentation coefficient can be corrected to the standard, $s_{20,w}$ value with Equation 5.11.

The presence of more than one sedimenting boundary can demonstrate either polydispersity or self-association phenomena. However, the converse is not necessarily true. A single sedimenting boundary is not in itself conclusive proof of sample homogeneity.[249] The downward slope in a plot of $S_{20,w}$ vs. concentration is a result of nonideality behavior.

5.1.7.2.7   Sedimentation Coefficient Distributions.   Stafford devised a method for obtaining sedimentation coefficient distributions from the time derivative of the concentration profile as well as new extrapolation procedures for reducing the effects of diffusion.[250-252] The differential sedimentation coefficient distribution function is defined as:[253]

$$\frac{d\{c(s)/c_o\}}{ds}\left(\frac{r}{r_m}\right)^2 \equiv g(s)_j \tag{5.14}$$

where j is "r" or "t" depending on whether it is calculated from dc/dr or dc/dt, respectively. The function g(s) ds gives the weight fraction of material sedimenting with sedimentation coefficients between s and s + ds.

The concentration distribution of the sedimenting material can be considered either as a function of radius at fixed particular instants of time or as a function of time at fixed specified values of radius. New methods for the analysis of sedimentation velocity data have extended the sensitivity of the UV scanning and Rayleigh interferometric optical systems of the analytical ultracentrifuge by 2 to 3 orders of magnitude.[238,254, 255]

Stafford analyzed the sedimentation boundary using the time derivative of the concentration profile and instrumental techniques that employ a rapid acquisition video-based Rayleigh optical system. Use of the time derivative achieves an automatic optical background correction. The video system allows signal averaging of the sedimentation patterns, resulting in a considerable increase in the signal-to-noise ratio. Sedimenting boundaries are represented as apparent sedimentation coefficient distribution functions, g(s*) vs. s*, where s* is the apparent sedimentation coefficient defined as $s = \ln(r/rm)/\omega^2 t$ and g(s*) has units proportional to concentration per Svedberg. A plot of g(s*) vs. s* is geometrically similar to the corresponding plot of dc/dr vs. r obtained with the schlieren optical system.[251]

At values of $\omega^2 st/D$ sufficiently large that any component's boundary in an ideal mixture has become separated from the others, the following relation was proposed by Stafford to obtain the molecular weight of that component:[255]

$$g^*(s)_{max} = \left(\frac{s}{2\pi D}\right)^{1/2}(\omega^3 r_m)\left(\frac{t}{\left(1-e^{-2\omega_{st}^2}\right)^{1/2}}\right) \tag{5.15}$$

The slope from a plot of g*(s)max vs. the last term in parenthesis in Equation 5.15 is proportional to the square root of the ratio s/D and, therefore, also proportional to the square root of the molecular weight through the Svedberg equation, so that

$$M = \frac{2\pi RT}{\left(1-\bar{v}\rho\right)\omega^6 r_m^2}(slope)^2 \tag{5.16}$$

where the symbols have their usual meaning.

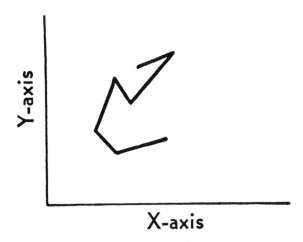

FIGURE 5.19 Brownian movement of a microparticle.

5.1.8 Diffusion Coefficient

In calculating the molecular weight from sedimentation velocity runs, an accurate estimate of the diffusion coefficient, D, is needed. However, the diffusion coefficient is probably the least frequently analyzed of the fundamental physical characteristics of viruses. The International Committee on Taxonomy of Viruses does not list D as a property of viruses in taxonomy.[35]

Classical determinations of D are, in general, difficult, time-consuming, and require large amounts of material. However, the use of dynamic light scattering procedures, involving the laser, make the determination of D easily obtained in a reasonable time period.

In this discussion, some of the classical methods are presented because D can be determined with the analytical ultracentrifuge. A discussion then follows whereby the use of a laser light scattering technique is described.

5.1.8.1 Diffusion

Dissolved substances diffuse from a region of higher concentration to a region of lower concentration. Diffusion is a spontaneous process and is closely related to Brownian motion, and we may infer that the particles of a substance diffuse because of their Brownian motion (Figure 5.19).

Two types of diffusion are recognized: translational and rotational. For the former, free diffusion is permitted to occur, or it may be modified by the application of an external force, such as centrifugal force in the sedimentation equilibrium technique. Rotational diffusion is characteristic of the type of molecule. A spherical molecule is characterized by a single rotary diffusion coefficient, an ellipsoid of revolution by two coefficients, one for each axis, and a general ellipsoid molecule by three coefficients, one corresponding to each of the axes. Rotary diffusion coefficients are measured by flow birefringence and, in the case of molecules that are dipolar, electrical methods.[256]

The tendency of a substance to diffuse may be expressed as its diffusion coefficient D, which is defined by Fick's first law. Fick stated that for diffusion in one direction the flow J of a substance through a plane perpendicular to the direction of diffusion is directly proportional to the rate at which concentration changes with distance, dc/dx, the concentration gradient.

$$J = -D\left(dc/dx\right) \tag{5.17}$$

The negative sign indicates that the flow is in the direction opposite to that in which c increases. If the flow J is expressed in g cm^{-2} s^{-1} and the concentration gradient in g cm^{-4}, the diffusion coefficient will have the units of cm^2 s^{-1}. Values for viruses lie, in general, in the range 0.2×10^{-7} to 2×10^{-7} cm^2/s.

Since D is concentration dependent, the value of D should be determined at a number of different initial concentrations, and extrapolated to D°, the limiting value as c approaches zero. Measured diffusion coefficients may also be corrected for temperature, T, and the viscosity, η, of the solvent:

$$D_{20,w} = D_{obs}\left(\frac{293.2}{T}\right)\left(\frac{\eta_{T,w}}{\eta_{20,w}}\right)\left(\frac{\eta_s}{\eta_w}\right)$$

(5.18)

Classical methods for determining D generally include the use of a Tiselius cell or the analytical ultracentrifuge. For particles such as viruses, boundary spreading, the usual observation criterion, is so slow that measurement of diffusion coefficients takes days to months. This extended time period may give rise to bacterial growth.

The Tiselius electrophoresis cell consists of two parts: an upper chamber which receives dialyzed solvent, and a lower chamber for the solution. The upper chamber is free to slide over the lower chamber, connecting pure solvent to the virus solution. A boundary forms representing the path of solute macromolecules and this is followed and recorded with time by schlieren optics or the Rayleigh interference system. The diffusion coefficients of the separated top, middle, and bottom components of squash mosaic virus were determined using a Tiselius cell in a Model H Spinco (now Beckman) apparatus at 2°C. The observed D values were 0.702, 0.743, and 0.726 × 10⁻⁷ cm²/s for the top, middle, and bottom components, respectively. These values, corrected to conditions of water at 20°C, are 1.28, 1.35, and 1.32 × 10⁻⁷ cm⁻²/s.[257]

The analytical ultracentrifuge can be used for measurement of the diffusion coefficient.[258] All three optical systems — schlieren, Rayleigh interference, and absorption — may be employed in observing the spread of the boundary with time. A commonly used method makes use of a synthetic boundary cell where dialyzed solvent in equilibrium with the solution is layered over the solution as the rotor reaches about 4000 to 6000 rpm. At this speed, the increased pressure of the solvent column is sufficient to force solvent through the narrow capillary between the sectors on to the surface of the solution. The boundary and meniscus are nearly vertical and in line with the optical axis. Scans of the cell contents at different times permit measurement of both the concentration in the plateau region, c_p, and the concentration gradient at the boundary, $(dc/dr)_b$, by numerical differentiation of the data. If the boundary is symmetrical, its position will be that of maximum concentration gradient, and will occur at the point where $c = c_p/2$. The diffusion coefficient is calculated as 4π times the slope of a plot of $[c_p/(dc/dr)]^2$ vs. time in seconds.

If the boundary created is sufficiently sharp the plot will pass through the origin. If the line cutting the time axis is away from the origin, a zero time correction, Δt, is applied. Creeth and Pain assert that for valid results $D\Delta t$ should be less than 10^{-4} cm².[259]

Chervenka[234] suggested an alternative method for analyzing data from the experiment described above. Using interference optics the values are recorded for c, $r_{1/4}$, and $r_{3/4}$: the concentration in the plateau, and the radial positions, where $c = c_p/4$ and $3c_p/4$, respectively (Table 5.5). If $\Delta r = r_{3/4} - r_{1/4}$, then D may be obtained from the slope of a plot of $(\Delta r)^2$ in cm² vs. time in seconds. From Equation 5.19, the slope of the plot multiplied by 0.275 (i.e., 1/3.64), is the apparent diffusion coefficient in cm²/s for the concentration of solute used in the experiment:

$$D = \frac{\Delta r^2}{4y^2 t} = \frac{\Delta r^2}{t}\frac{1}{3.64}$$

(5.19)

The value 0.275 comes from the width of a Gaussian distribution between 1/4 and 3/4 levels.[260]

To determine D_o, the diffusion coefficient at infinite dilution, the experiment is repeated using a series of solute concentrations and the diffusion coefficients extrapolated to zero concentration. To determine $D_{20,w}$, Equation 5.18 is used with appropriate values for the viscosity of the solvent and water.

TABLE 5.5   Data for Calculation of Diffusion Coefficient

Frame No.	Time (sec)	J (fringes)	J/4 (fringes)	$x_{1/4}$ (cm)	3J/4 (fringes)	$x_{3/4}$ (cm)	Δx (cm)	Δr	Δr^2
1	430	16.92	4.23	14.512	12.69	14.599	0.087	0.0404	1.63×10^{-3}
2	910	16.90	4.23	14.500	12.68	14.621	0.121	0.0563	3.17×10^{-3}
3	1,390	16.88	4.22	14.486	12.66	14.635	0.149	0.0694	4.81×10^{-3}
4	1,870	16.87	4.22	14.478	12.66	14.648	0.170	0.0791	6.26×10^{-3}
5	2,350	16.86	4.22	14.467	12.66	14.658	0.191	0.0887	7.86×10^{-3}
6	2,830	16.87	4.22	14.460	12.66	14.668	0.208	0.0966	9.31×10^{-3}
7	3,310	16.83	4.21	14.453	12.63	14.678	0.225	0.1045	10.92×10^{-3}
8	3,790	16.84	4.21	14.447	12.63	14.687	0.240	0.1115	12.43×10^{-3}

From Chervenka, C.H., *A Manual of Methods for the Analytical Ultracentrifuge,* Beckman Instruments, Palo Alto, CA, 1969. With permission.

5.1.8.2   Measurement of D by Dynamic Light Scattering (DLS)[261]

The technique of dynamic light scattering (DLS) allows the diffusion coefficients of very large particles such as viruses and DNA to be measured in a matter of minutes.

The frequency dependence of scattered light on dynamic molecules was predicted and observed in the 1920s in experiments utilizing Raman and Brillouin scattering. In 1926, Mandelshtam demonstrated that translational diffusion coefficients of polymers could be obtained from the spectrum of their scattered light. With the advent of the laser in the early 1960s, monochromatic, coherent light could be utilized and application to macromolecular analyses began. DLS has a number of synonyms:[262] intensity fluctuation spectroscopy (IFS), light-beating spectroscopy, photon correlation spectroscopy (PCS), or quasielastic light scattering (QLS). Thus, in 1964 Pecora[263] developed a theory to explain the spectrum of scattered light in terms of the motion of macromolecular scatterers. Also, at this time, Cummins et al.[264] measured the diffusion coefficient of polystyrene latex spheres from the spectral broadening of the scattered light. For biological macromolecules, Dubin et al.[265] in 1967 made the first application of DLS to the measurement of diffusion coefficients of proteins, TMV, and DNA. In 1970, Dubin and coworkers obtained accurate diffusion coefficients for coliphages T4, T5, T7, and λ phage, and with values of the sedimentation coefficients and partial specific volumes, computed the molecular weight for these viruses.[266] For the determination of D, all experimental runs were performed within 1 h, except in the case of T7. This phage had to stand for about 10 h after dilution, in order to give valid results.

Bloomfield and Lim have written a detailed review on dynamic light scattering.[261] Descriptions of the method are presented here as treated by Markham[267] and Dubin et al.[266]

Brownian motion causes local concentration differences and local variations in the refractive indices of solutions. A monochromatic beam of light is frequency modulated by these local refractive index variations, and the extent of the modulation can be used to determine the local fluctuation and then D. By using the extreme monochromaticity of a laser, the frequency variations may be detected by self-heterodying in the audio frequency range. In this connection, a scattering cell cuvette is illuminated by a laser beam, and the light scattered at an appropriate angle is detected by a photomultiplier. The output is passed through a narrow-band audio-spectrum analyzer and then through an analog squarer to give the effective power at each wavelength, which is plotted on a chart.

In measuring the diffusion coefficient of several phages, Dubin et al.[266] measured the spectrum of light scattered by these viruses. The spectral width is directly proportional to the diffusion coefficient of the scattering macromolecules. An ultra-high-resolution spectrometer was used to observe the spectrum of scattered light.

The 6328 Å radiation of a Spectra-Physics model no. 125 laser (~50 mw) was focused in the scattering cell — a standard spectrometric cuvette. The transmitted laser beam was monitored by a silicon solar cell which controlled a servo to ensure constant laser power. Light scattered at an angle θ was focused onto an RCA 7265 photomultiplier tube, the output of which was analyzed by a General Radio 1900A audio-spectrum analyzer. The output of this device is proportional to the voltage rather than the power spectrum and its output was squared before display on the strip chart recorder. By slowly sweeping the center frequency of the spectrum analyzer the entire self-beat spectrum was mapped out. About 1 h was required to obtain the complete spectrum. The requirement of precise temperature stability in conventional determinations of D was not necessary with the spectrometer since, to first order, convection currents do not affect this measurement of D.[265]

Light of wave vector k_o is allowed to fall on an assemblage of macromolecules, each of which, at some time t, is located at the position $r_j(t)$. The intensity of the light scattered with wave vector k is determined at each instant of time by the superposition of the phases of the waves scattered by each of the molecules at the various positions $r_j(t)$. The intensity of the scattered light fluctuates randomly about the value of the mean intensity because the phase of the light waves scattered by each molecule is constantly changing as the particles behave randomly — undergo a random walk. The phase of the wave scattered by a particle at r_j relative to a point at the origin (O) is seen to be $\phi_j = (k_o - k) \cdot r_j(t) \equiv K \cdot r_j(t)$. The difference $K = k_o - k$ is defined as the scattering vector, whose magnitude is

$$K = \frac{4\pi}{\left(\lambda/n\right)} \sin\left(\frac{\theta}{2}\right) \qquad (5.20)$$

where θ is the angle between the wave vectors of the incident and scattered light (k_o and k), n is the index of refraction of the solution, and λ is the wavelength of the incident light in vacuum. Because of the continually changing arrangement of particles the intensity I(t) of the scattered light fluctuates randomly. This random variable I(t) can be characterized by its spectral power density or by its correlation function. If t_c is the "correlation time" beyond which intensity fluctuations at t are uncorrelated with those at t + r, then the width of the spectrum $[(\Delta\omega)_{1/2}]$ of the intensity fluctuations is related to r_c:

$$\left(\Delta\omega\right)_{1/2} = \frac{1}{r_c} \qquad (5.21)$$

where $(\Delta\omega)_{1/2}$ is the half-width at half-maximum in radians per second. The spectral width of the scattered light intensity is found by estimating the magnitude of r_c: the correlation time for the random fluctuations in I(t). The spectral width of the intensity fluctuations has been shown to be related to D, the translational diffusion coefficient, by the equation

$$\Delta v_{1/2} = \left(\frac{2}{2\pi}\right) DK^2 \qquad (5.22)$$

where $\Delta v_{1/2}$ is the half-width at half-maximum in Hertz.

Detailed calculations also showed that if the molecule is small compared to the light wavelength and undergoes isotropic random walk, the line shape is Lorentzian. In the experiments conducted by Dubin et al.,[266] the line shapes were found to be accurately Lorentzian. The diffusion coefficient D can be deduced unambiguously from the line width, since the scattering vector K is known from Equation 5.20 from the scattering angle θ and the index of refraction of the solution.

Presently, instead of a wavelength analyzer, short-time fluctuations in intensity, caused by the movements of the macromolecules, are measured.[262] These changes in intensity, or numbers of photons received by a detector, are recorded using a computer, designed for such measurements — an "autocorrelator." From suitable analysis of the change of the "autocorrelation function," with time, translational diffusional information can be obtained.

The correlator calculates the normalized intensity correlation function $g_{(\tau)}^{(2)}$ as a function of the delay time τ:

$$g_{(\tau)}^{(2)} = \left[\left\langle I(t) \cdot I(+\tau) \right\rangle \Big/ \left\langle I \right\rangle^2 \right]$$ (5.23)

The angular brackets indicate that the products are averaged over long times (which can be of the order of 1 minute or higher depending on the size of the scattering particles and the power of the laser) compared with τ.

For dilute Brownian systems, macromolecules and macromolecular assemblies with $M \leq 100 \times 10^6$, which are also quasispherical, the normalized autocorrelation function $g_{(\tau)}^{(2)}$ is related to the translational diffusional coefficient, D, by:

$$\left[g_{(\tau)}^{(2)} - 1 \right] = e^{-Dk^2\tau}$$ (5.24)

where k is defined by Equation 5.20; n is the refractive index of the medium, and λ the wavelength of the incident laser light. D can be found from a plot of Ln $[g_{(\tau)}^{(2)} - 1]$ vs. τ.

The translational diffusion coefficient so obtained will be a function of solvent conditions. As with the sedimentation coefficient, it is usual to correct the value to standard conditions (water at 20°C), to give $D_{20,w}$. This corrected value at a finite concentration will be an apparent one, and so measurement at several concentrations and extrapolation to zero concentration to give a value at infinite dilution $D_{20,w}^0$ is normally necessary.

Equation 5.24 is only exact for spherical particles. For nonspheroidal macromolecular scatterers, the contribution from rotational diffusional effects may not be negligible at higher angles, and the measured translational diffusion coefficient will be an apparent value with respect to angle. Therefore, in addition to measurent of $D_{20,w}$ as a function of concentration and extrapolation to zero concentration, a similar set of measurements as a function of angle and extraploation to zero angle are necessary for asymmetric scatterers. These two extrapolations can be done on the same set of axes to give a "dynamic Zimm plot"[268] (see Chapter 6, Section 6.3). If the sample is polydisperse or self-associating, the logarithmic plot will tend to be curved, and the corresponding diffusion coefficient will be a z-average. The z-average $D_{20,w}^0$, when combined with the weight average $s_{20,w}^0$ in the Svedberg equation, gives a weight average molecular weight, M_w.

As with classical light scattering, the presence of dust in the sample remains a problem. The technique is best suited for macromolecules with molecular weight $\geq$ 100,000. Measurement at a single angle, conventionally 90°, gives insufficient information to obtain D for asymmetric scatterers. Extrapolation to zero angle is necessary, which can cause problems, since at low angles the dust problem is at its greatest.

A description of how classical light scattering equipment can be modified for dynamic work is contained in a report by Godfrey et al.[269] Figure 5.20 shows a modern instrument for dynamic light scattering. Most of the dynamic light scattering instrumentation also facilitates total intensity (i.e., "classical") light scattering measurements.

With respect to viruses, in addition to obtaining the diffusion coefficient, investigations of changes in these macromolecules can also be studied by dynamic light scattering.[270,271]

For a heterogeneous system, it is also possible to obtain a parameter that indicates the spread of diffusion coefficients — the normalized z-average variance of the diffusion coefficients, referred to as the "polydispersity factor." It is possible to relate this value to the distribution of molecular weight.[272] Commercial software, such as CONTIN, is available for inverting the autocorrelation data directly to give distributions of diffusion coefficient and equivalently particle size.[273,274] In addition, online coupling of dynamic light scattering to gel permeation chromatography has also been demonstrated.[275]

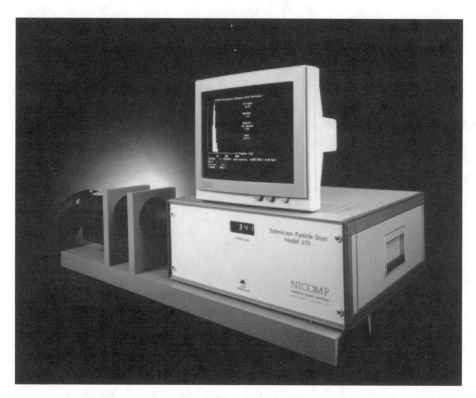

FIGURE 5.20   NICOMP 370/VHPL Submicron Particle Sizer. The instrument, outfitted with a 75 mW air-cooled argon-ion laser, provides light scattering data including a value for the translational diffusion coefficient, D, used in the calculation of molecular weight. (Courtesy: Particle Sizing Systems (PSS), Santa Barbara, CA.)

5.1.8.3   Alternative Procedures for Obtaining D and S: DLS Applied to Density Gradients

Density gradient centrifugation may be combined with DLS to obtain both S and D from a preparative ultracentrifuge tube. Combination of S and D then yields the molecular weight of the solute by Equation 5.13. Koppel[276] has used this procedure to characterize *Escherichia coli* ribosomes and their 30S and 50S subunits. The sedimentation bands were stabilized by a sucrose gradient. Upon attainment of adequate band separation, the tube was removed from the ultracentrifuge and prepared for scanning by the DLS laser beam. Two scans were made. From the first scan, the average intensity of the scattered light was measured as a function of height in the tube. This procedure permitted location of the bands. The distance moved by each band, divided by the time of centrifugation and the angular acceleration, $\omega^2 r$, gave the sedimentation coefficient. The magnitude of the scattered intensity gives a measure of the particle concentration at each point. The solvent refractive index was obtained from tables relating sucrose concentration and refractive index.

In the second scan along the tube, the autocorrelation function was measured at each point, permitting the determination of the diffusion coefficient of the macromolecules in each band. To correct S and D to standard conditions of $S_{20,w}$ and $D_{20,w}$, respectively, an accurate calibration of the sucrose gradient is needed. The density is also required for standardization to $s_{20,w}$. The correlation of sucrose concentration with refractive index, viscosity, and density is available from tables. To characterize the gradient, two methods may be used. The refractive index of the solution collected from several positions in the cell can be measured by differential refractometry. Alternatively, the diffusion coefficient of a standard scattering sample, such as polystyrene latex spheres, 91.0 nm in diameter, can be measured at varying heights in the sucrose gradient. From the measured diffusion coefficient and known radius, the viscosity of the sucrose solution at each height can be determined and, thereby, the solution density from tables.

TABLE 5.6 Diffusion Coefficients (D) of
Viruses from Dynamic Light Scattering Studies

Virus	D*	Ref.
Avian myeloblastosis	0.268	278
Vesicular stomatitis	0.29	279
Rous sarcoma	0.291	278
Tobacco mosaic	0.39	280
Infectious pancreatic necrosis	0.67	281
Turnip yellow	1.44	281
R17	1.534	282
QB	1.423	282
Tomato bushy stunt	1.246	282
PM2	0.650	282
T2	0.644	282

* $D = 10^{-7}$ cm^2/s.

Once the sedimentation and diffusion coefficients and the density have been determined, the molecular weight of the solute in each band can be evaluated.

Using similar procedures involving sedimentation in capillary tubes, Lowenstein and Birnboim[277] measured the sedimentation and diffusion coefficients and molecular weights of R 17 bacteriophage. Table 5.6 gives values of D of viruses by dynamic light scattering.

5.1.9 Sedimentation Equilibrium

The sedimentation equilibrium method studies the concentration distribution of molecules at equilibrium positions. The concentration distribution can be considered kinetically or thermodynamically. Basic equations for both approaches were derived originally by Svedberg and Pedersen.[224]

5.1.9.1 Theory

The kinetic description of sedimentation equilibrium treats equilibrium as a state in which transport in a centrifugal direction is completely counterbalanced by transport in a centripetal direction. While the kinetic description of equilibrium provides a good empirical definition of the equilibrium state in the cell, it does not provide a strong foundation for a general and rigorous theory of equilibrium. Thus, many of the theoretical uncertainties inherent in the sedimentation velocity method are also inherent in the kinetic (transport) description of sedimentation equilibrium.[283,284]

Current theory of sedimentation equilibrium stems from studies by Goldberg,[285] who introduced the concept that the equilibrium distribution of solute should be considered solely in thermodynamic terms, rather than as a state reflecting a balance between fluxes owing to sedimentation and diffusion.[286]

Thermodynamic equilibrium can be defined as a state in which the total potential of any component in the centrifuged solution is constant. The escaping tendency of each component at any point in the cell is equal to its escaping tendency at any other point, and the temperature is uniform throughout the cell. The total potential is the sum of the solution's chemical and centrifugal potentials. The chemical potential is a function of concentration and pressure. The centrifugal potential is a function of the centrifugal field and position in the cell.

If appropriate expressions are supplied for the effect of concentration (RT/c) pressure (Mv), and distance ($-M\omega^2 x$) and if the derivative of the total potential is set to zero, the following equation results:

$$M = \frac{2RT}{\left(1 - \bar{v}\rho\right)\omega^2} \frac{d\ln c}{d\left(x^2\right)} \qquad (5.25)$$

where M = molecular weight, R = the gas constant (8.313×10^7 ergs/degree mole), T = absolute temperature, $\bar{v}$ = partial specific volume, ρ = density of solution, ω = angular velocity, c = concentration, x = distance from the axis of rotation.

This is the second fundamental equation in sedimentation analysis. It is the same equation obtained if Svedberg's equation for two-component systems, $M = RTs/D (1-\bar{v}p)$ is combined with the equation yielding the net transport across a surface at a distance x.

The term "molecular weight of a solute" is a definite term only if the substance is composed of one molecular species. If the solution is composed of more than one molecular species, the molecular weight term must be replaced by an average molecular weight. The common indexes used to express the statistic average are number (n), and weight (w or z). These averages, called uniform averages, appear only if the molecular weight is calculated from experimental values that are a function of molecular weight alone, otherwise mixed averages result.

M_w, the weight-average molecular weight and M_z, the z-average molecular weight, can be readily evaluated from sedimentation equilibrium experiments. M_n, the number-average molecular weight, can also be determined, but with more difficulty than that involved when determining the other values. These molecular weight averages are defined in the following equations, where n is the number of molecules of molecular weight M:[287]

$$M_n = \frac{\Sigma n_i M_i}{\Sigma n_i} = \frac{\Sigma c_i}{\Sigma c_i / M_i} \qquad (5.26)$$

$$M_w = \frac{\Sigma n_i M_i^2}{\Sigma n_i M_i} = \frac{\Sigma c_i M_i}{\Sigma c_i} \qquad (5.27)$$

$$M_z = \frac{\Sigma n_i M_i^3}{\Sigma n_i M_i^2} = \frac{\Sigma c_i M_i^2}{\Sigma c_i M_i} \qquad (5.28)$$

Equation 5.25 refers to a two-component ideal solution, and variations in density from both concentration and pressure changes in the cell are usually neglected.

Fujita referred to a solution as "ultracentrifugally ideal" when all cross-term diffusion coefficients equal zero and when s, the sedimentation coefficient, and D, the major diffusion coefficient, can be considered constant.[288]

Goldberg described a system at equilibrium in terms of a continuous sequence of phases of fixed volume and of infinitesimal depth in the direction of the centrifugal field. Allowing the restriction that the total potential and the temperature are uniform, he obtained the following equation:[285]

$$M_i\left(1 - \bar{v}_i^{(x)} \rho^{(x)}\right) \omega^2 x dx = \sum_{k=1}^{r} \left(\partial \mu_i / \partial c_k\right)^{x_{T,P,c_j}} dc_k^{(x)} \qquad (5.29)$$

$$i = 0, 1 \ldots r$$

where μ_i = the chemical potential.

The solute species are denoted by the subscripts 1, 2 … r, and their concentrations expressed as mass per unit volume. Many terms in Equation 5.29 are written with x-superscript to indicate that the terms correspond to the surface at a distance x from the axis of rotation. The dependency of these terms on x results from the pressure gradient present in the cell and from the compressibility of the solute and solution.

One should note that the molecular weight of any component depends on all components in the system expressed by the terms $(\partial\mu_i/\partial c_k)$. Also, the density, ρ, which is introduced through the dependence on pressure, is the density of the solution and is a function of position in the cell.

Goldberg treated nonideal systems by expressing the chemical potential (μ_i) in terms of both concentration and the activity coefficient (γ) of the solute:

$$M_i\left(1-\bar{v}_i^{(x)}\rho^{(x)}\right)\omega^2xdx = RTd\ln c_i^{(x)} + RT\sum_{k=1}^{r}\left(\frac{\partial\ln\gamma_i}{\partial c_k}\right)_{T,P,c_j}^{(x)}dc_k^{(x)} \qquad (5.30)$$

Equation 5.30 is basic to all sedimentation equilibrium studies. The equation, or some approximation of it, is used in the interpretation of all experimental data. Some systems involve homogeneous solutes, which may or may not exhibit a concentration dependence, and other systems are concerned with ideal solutions of polydisperse materials. Between these two extremes are systems that exhibit not only a dependence on concentration but also possess solutes that are polydisperse.

For a two-component system containing a homogeneous solute and solvent, Equation 5.30 can be reduced to the following form:

$$M\left(1-\bar{v}^{(x)}\rho^{(x)}\right)\omega^2x = \frac{RT}{c}\frac{dc}{dx}\left(1+c\frac{\partial\ln\gamma}{\partial c}\right) \qquad (5.31)$$

The equation is written for the solute and usually disregards the dependence of partial specific volume and density upon position in the cell. Many protein solutions appear to exhibit ideal behavior and for such systems the term in the parenthesis to the right equals unity.

Generally, in sedimentation equilibrium experiments, an integrated form of Equation 5.25 is used for determining molecular weight.

A plot of ln (concentration) vs. (radius)^2 for a single, ideal solute at sedimentation equilibrium yields a slope proportional to the buoyant molecular weight, M $(1-\bar{v}\rho)$.[289] Alternatively, the data of c vs. r^2 can be fitted by least squares to give the best estimate of $M(1-\bar{v}\rho)$. The value for M can be calculated from the partial specific volume of the solute and the density of the solution.

In a practical sense final thermodynamic equilibrium is never actually achieved because of the requirement of infinite time. The practical equilibrium condition is achieved in finite time, and for computational purposes is the same as the theoretical equilibrium state. The practical equilibrium condition, which is the subject in the discussion to follow, is that in which the concentration distribution in the centrifuge cell does not change with time within the experimental constraints needed to measure the distribution.[234]

The sedimentation equilibrium technique has certain advantages over the sedimentation-diffusion procedure. It has a secure theoretical thermodynamic basis so that fewer assumptions are required to make a practical application of the method to experimental data; it is simpler experimentally, and it is more accurate. One prime disadvantage in conventional equilibrium experiments is that more time is needed for the use of the ultracentrifuge.

5.1.9.1.1 Experimental Considerations.[290] As for sedimentation velocity studies, one should work, if possible, with a solvent of a sufficiently high ionic strength, to provide adequate suppression of charge effects. Such effects contribute to the thermodynamic nonideality of the system.

For sedimentation equilibrium studies, it is advisable to dialyze the solution against the solvent prior to a sedimentation equilibrium run, to avoid possible redistribution phenomena of the aqueous solvent components (salt ions, etc.) themselves.

For large macromolecular systems, nonideality may be significant at the concentrations used, and some form of correction for thermodynamic nonideality effects may be necessary. The measured molecular weights at a finite concentration will be "apparent" molecular weights. Because of the instability of rotor systems at very low speeds, the technique may be unsuitable for large viruses having a molecular weight of greater than 20 million.

Rayleigh interference optics usually provide the best optical records for molecular weight analysis. The absorption optical system is, however, the most convenient.

5.1.9.1.2   Length of Run.   Smaller molecules get to sedimentation equilibrium faster than larger ones. For molecules of M greater than 10,000, less than 24 h are required. Large macromolecules take 48 to 72 h, although time to equilibrium can be decreased by "overspeeding," running at a higher speed for a few hours before setting to the final equilibrium speed. Shorter columns, as low as 0.5 mm, may be used, although the accuracy of the molecular weights will be lower. The short column method offers the advantage of fast equilibrium, usually less than 24 h.

5.1.9.1.3   Data Capture and Analysis.   If scanning absorption optics are used, the equilibrium patterns can be digitized directly online into a microcomputer or off-line via a graphics digitizing pattern. The average slope of a plot of ln (absorbance) vs. r^2, the square of the radial distance from the center of the rotor, will yield the weight average molecular weights (Equation 5.25).

$$M = \left(dLnA / dr^2\right) \times 2RT / \left(1 - \bar{v}\rho\right)\omega^2$$

where $\bar{v}$ is the partial specific volume, ω is the angular velocity (rad/s), and ρ the solution density. In general, the solvent density ρ can be used instead without giving serious error in M.

With Rayleigh interference optics, the corresponding records of fringe displacement, J, vs. radial displacement, r, can be obtained either manually using microcomparators, off-line using a laser densitometer,[291] or directly online into a microcomputer.[292] An average slope of a plot of Ln{J} vs. r^2 can be used in much the same way as Ln{A} vs. r^2, yielding the weight average molecular weight. If schlieren optics are used, the average slope of a plot of Ln {(1/r) dn/dr} vs. r^2, where dn/dr is the refractive index gradient at a given radial position, r, yields the z-average molecular weight.

For larger viruses, nonideality may become significant, and this will tend to cause downward curvature in the Ln{A} or Ln{J} vs. r^2 plots. If the material is not significantly heterogeneous, then a simple extrapolation from a single experiment of point (apparent) molecular weight to zero concentration can be made, to give the infinite dilution, "ideal," value. In general, reciprocals are usually plotted.[293]

5.1.9.2   Conventional Sedimentation Equilibrium Method

As noted by Chervenka,[234] the term "conventional" is used to distinguish procedures whereby the cell is centrifuged at rotor speeds sufficiently low to effect sedimentation equilibrium of the entire contents of the cell. Sedimentation proceeds in a cell with a fluid content of 3 to 5 mm, and the solute concentration at the meniscus is not zero. Methods that are not conventional are those concerned with decreasing the equilibrium time: short column, meniscus depletion, and approach-to-equilibrium procedures.

For all methods concerned with equilibrium, in order to compute the molecular weight, the concentration of solute throughout the cell must be determined when the condition of equilibrium has been reached. In addition, the rotor speed and temperature, the partial specific volume of the solute, and the density of the solution in the cell must be accurately known.

Two of the optical procedures discussed above are generally used in equilibrium experiments: absorption optics with the split-beam photoelectric scanning system and interference optics. In absorption optics, concentration data are most conveniently determined from the absorbance of the cell contents. While interference optics yields results of high accuracy, the concentration data are only relative, since absolute concentrations are not directly obtainable. The determination of fringe shifts near the ends of the cell, the meniscus and the bottom, requires an uncertain approximation. Moreover, in interference optics the use of the conventional synthetic boundary cell to obtain the required initial concentration may be subject to possible error.

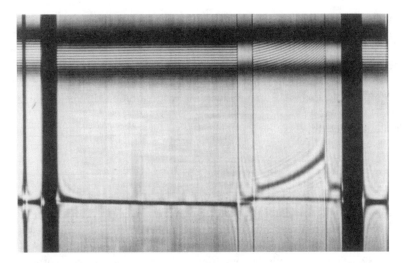

FIGURE 5.21 Combined schlieren (bottom) and Rayleigh interference optics (top) at sedimentation equilibrium. (Courtesy: Beckman Instruments, Palo Alto, CA.)

The schlieren system is less commonly used for equilibrium experiments, because it is considered to be less accurate than the interference system. However, it is used in the approach-to-equilibrium method. The variable sensitivity of the schlieren system makes possible its use for solute concentrations much higher than possible with interference optics. In addition, schlieren optics can be used with other optical systems to monitor the progress of the run, as is demonstrated in Figure 5.21.

5.1.9.2.1 Use of the Photoelectric Scanner.[233] One major advantage of using the photoelectric scanner is that very low concentrations of solute can be used. This facility allows a more accurate extrapolation to infinite dilution. Another advantage is that the concentration data required are obtained directly from the recorded data, assuming that the optical density of the solution is a linear function of concentration. A synthetic boundary run is not required and mass conservation need not be a concern.

Using a double sector scanner cell the optical density is read as a function of radial position at a series of points across the portion of the chart corresponding to the cell. The logarithm of each value of O.D. is determined and a plot of log O.D. vs. r^2 is made, the slope of which is proportional to molecular weight. The slope value is substituted in Equation 5.25 to compute molecular weight.

Manually, the accuracy of molecular weight values obtained by this technique depends upon the selection of the correct baseline position on the chart. After the equilibrium condition has been reached and recorded, the rotor is accelerated to the maximum allowable speed and run for a few hours. During this time the upper portion of the cell becomes depleted of solute by sedimentation. The rotor is then decelerated to the original equilibrium speed and further scans are made. The average position of the trace in the depleted region of the cell indicates the baseline for the entire cell. Pen deflection readings on the original equilibrium tracing are made from this position and can be used directly as absolute measures of concentration.

A conventional equilibrium run may make use of *overspeeding*[234] as a means to shorten the transient time, the time required to attain equilibrium conditions. The rotor is initially run at a speed higher than the equilibrium speed selected. During this period, some solute is transported by sedimentation toward the bottom of the cell. The rotor is then decelerated to a speed lower than the equilibrium speed to allow the diffusion of the sharp gradients formed during the initial period. Finally, the rotor is accelerated to the equilibrium speed, and allowed to run until the distribution of solute does not change.

Chervenka[234] has described in stepwise fashion the manual calculation of results of a run monitored by interference optics. The total fringe shift, J, is determined for each of the synthetic boundary exposures and a plot of J vs. run time is made. The plot is extrapolated to zero time, and the number of fringes is recorded as c_o, the initial concentration. The radial distance to the upper sample meniscus is measured, and then the radial distance to the center of each fringe. These measurements are made by starting at the meniscus on a dark fringe and making the first reading on the next bright fringe. The position of each bright fringe is read to the bottom of the cell. Next, the position of the last dark fringe and the position of the bottom meniscus are read. A fraction of a fringe generally remains between the last dark fringe and the bottom meniscus. This fraction is determined as indicated in the discussion of Table 5.7, in which r^2 is computed for each measured position. The concentration at the meniscus, c_m, is obtained from Equation 5.32.

The required integral of this equation is estimated as

$$\Delta j \sum_{r_m}^{r_b} r^2$$

(where $\Delta j = 1.0$) by summing the values of r^2 for each bright fringe, plus a contribution for the fractional fringe. The contribution is determined by averaging the r^2 values for the last whole fringe and the bottom of the cell, and multiplying by the fractional fringe number. The concentration at the meniscus, c_m, is determined from Equation 5.32, using the total fringe shift as $c_b - c_m$. The concentration in fringes at various radial positions in the cell is determined by adding the fringe shift at each radial position to the concentration at the meniscus. A plot is made of $\log_{10} c$ vs. r^2 (Figure 5.22) and the slope of the plot, d $\log c/dr^2$, is substituted into Equation 5.25 to calculate the molecular weight, after values for $\bar{v}$ and ρ have been determined.

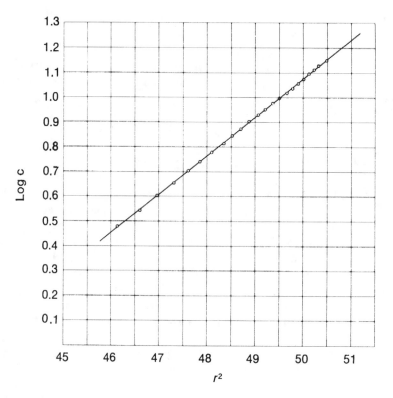

FIGURE 5.22 Log c vs. r^2 plot used for the analysis of sedimentation equilibrium data. (Courtesy: Beckman Instruments, Palo Alto, CA.)

TABLE 5.7 Sample Data for Calculation of Molecular Weight
by Conventional Sedimentation Equilibrium

Fringe No.	x Scale Reading (cm)	r (cm)	r^2	c (fringes)	$\log_{10}c$
0 (meniscus)	14.196	6.792	46.131	3.02	0.480
0.5	14.268	6.827	46.408	3.52	0.547
1.5	14.377	6.879	47.321	4.52	0.655
2.5	14.457	6.917	47.845	5.52	0.742
3.5	14.531	6.953	48.344	6.52	0.814
4.5	14.587	6.979	48.706	7.52	0.876
5.5	14.639	7.004	49.056	8.52	0.930
6.5	14.684	7.026	49.365	9.52	0.979
7.5	14.726	7.046	49.646	10.51	1.001
8.5	14.759	7.062	49.872	11.52	1.061
9.5	14.796	7.079	50.112	12.52	1.098
10.5	14.824	7.093	50.311	13.52	1.131
11.15 (bottom)	14.850	7.105	50.481	14.17	1.151

Note: Last dark fringe is number 11.0. Fractional fringe = 0.15.
From Chervenka, C.H., *A Manual of Methods for the Analytical Ultracentrifuge*, Beckman Instruments, Palo Alto, CA, 1969. With permission.

$$c_m = c_o - \frac{r_b^2(c_b - c_m) - \int_{r_m}^{r_b} r^2\, dc}{r_b^2 - r_m^2}$$

$$= c_o - \frac{r_b^2(c_b - c_m) - \Delta j \sum_{r_m}^{r_b} r^2}{r_b^2 - r_m^2}$$

(5.32)

An example of the calculation of molecular weight by conventional sedimentation equilibrium is as follows:[234]

$$(1 - \bar{v}p) = 0.277,\ \text{rotor speed} = 15{,}000\ \text{rpm, temperature} = 12.4°C,$$
initial concentration = 7.23 fringes.

From Table 5.7, the value obtained from the required integral = 544.545. Using Equation 5.32:

$$C_m = 7.23 - 50.481\ (11.15 - 0) - 544.545/50.481 - 46.131 = 3.02$$

From Figure 5.22 (log c vs. r^2 plot),

$$d\log c/d(r^2) = 0.156$$

Using Equation 5.25

$$RT/(1 - \bar{v}p)\ \omega^2 = (8.313 \times 10^7)\ (285.6/0.277)\ (2.467 \times 10^6) = 3.474 \times 10^4$$

$$M = 2(3.474 \times 10^4) \times 2.303 \times 0.1565 = 2.504 \times 10^4$$

The procedure described, when carried out at a single concentration, yields only an apparent value for molecular weight. To determine the molecular weight at infinite dilution — zero concentration — the run is repeated with other concentrations of the solute in the same solvent, including the lowest concentration possible. Extrapolation to zero concentration is made, where ideality is assumed.

There are alternative methods to that described. Lansing and Kraemer[287] provided a method that yields the apparent molecular weight of a homogeneous solute or the apparent weight average molecular weight of a polydisperse solute. Equation 5.33 gives a calculation of the molecular weight using the same experimental procedure described above. From a fringe pattern of the cell contents at sedimentation equilibrium, the total fringe shift from the upper meniscus of the sample solution to the bottom is determined as $c_b - c_m$. This value, together with c_o, determined from the fringe shift in the synthetic boundary cell, is used to calculate M:[294]

$$M = \frac{2RT}{\left(1 - \bar{v}\rho\right)\omega^2\left(r_b^2 - r_m^2\right)} \frac{c_b - c_m}{c_o} \tag{5.33}$$

This method allows a quick analysis of the results. Its disadvantages are that no information concerning the molecular weight distribution is obtained, and the results rely entirely upon the fringe counts at the ends of the fluid column, where there may be considerable uncertainty.

5.1.9.3 Meniscus Depletion Sedimentation Equilibrium Method[295,296]

This technique requires a speed of two to three times that used for conventional sedimentation equilibrium. At such speeds, all macromolecular solute is sedimented out of the region of the cell near the meniscus. The transient time is shorter because of the relatively high speed used.

Interference optics are used for this technique because the concentration data required are obtained directly and accurately. The interference fringe count through the cell image becomes a direct measure of the absolute concentration of solute to any position in the cell. A synthetic boundary run to determine the initial concentration is not required.

Molecular weights close to infinite dilution values can be obtained directly. In a single run accurate values can be determined over a wide range of concentrations down to about 0.03 mg/ml in a 30 mm cell.

A disadvantage of the method is that the conditions of zero concentration at the meniscus may not be achieved sufficiently because of back-stirring caused by the instability of the solution in that region of the cell, or because of slight vibrations in the rotor.

Near the end of the estimated run time, the pattern should look like the fringes shown in Figure 5.23. Interference photos are taken at 1 h intervals until there is no measurable difference in the patterns. At equilibrium the fringes should be straight over approximately the upper half of the cell to ensure that there is zero concentration of the solute at the meniscus.

With an equilibrium plate in the comparator, the horizontal cross hairs are set parallel to the fringes in the reference hole images. At the meniscus, the vertical position of a clear fringe is determined at intervals of 0.05 cm. These readings are repeated at identical radial positions on the initial fringe photo and then the scale readings are subtracted from the corresponding readings for the equilibrium photo. The results are recorded as $y(r)$, as in Table 5.8. The $y(r)$ values for the linear portion of the equilibrium photo are averaged. This average, y_o, is subtracted from the remaining values of $y(r)$, to yield the fringe displacement, $[y(r) - y_o]$ for each of the radial intervals. The distance to the center of rotation, r, is determined for each interval, and is squared. A plot is made of log $[y(r) - y_o]$ vs. r^2 as in Figure 5.24. The slope of the plot is proportional to the apparent molecular weight of the solute. Equation 5.25 is used to compute molecular weight.

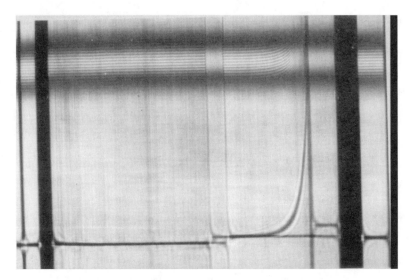

FIGURE 5.23 Combined schlieren (bottom) and Rayleigh interference optics (top) at sedimentation equilibrium with the meniscus depletion procedure. (Courtesy: Beckman Instruments, Palo Alto, CA.)

An example of the calculation of molecular weight by meniscus deletion sedimentation equilibrium[234] is as follows:

$$(1-\bar{v}p) = 0.262$$

$$\text{rotor speed} = 30,000 \text{ rpm}$$

$$\text{temperature} = 16.8°C$$

From Figure 5.24, log $[y(r) - y_o]$ vs. r^2 plot:

$$\frac{d \ \log\left[y(r)-y_o\right]}{d\left(r^2\right)} = 0.659 \left[\text{used as } d \log c / d\left(r^2\right)\right]$$

Using Equation 5.25

$$\frac{RT}{\left(1-\bar{v}p\right)\omega^2} = \frac{\left(8.313 \times 10^7\right) \ 290.0}{0.262 \ \left(9.870 \times 10^6\right)} = 9.323 \times 10^3$$

$$M = 2\left(9.323 \times 10^3\right) \times 2.303 \times 0.659 = 2.83 \times 10^4$$

5.1.9.4 Short Column Sedimentation Equilibrium Method

This method makes use of the fact that the time required to reach equilibrium is a function of the square of the fluid column height in the centrifuge cell.[297] A reduction in the volume of sample used leads to greatly reduced operating times. Pederson noted this principle when the theory of sedimentation equilibrium was being elucidated.[224]

The short column method is defined as that in which the sample column is 1 mm or less. Van Holde and Baldwin described sedimentation equilibrium experiments using short columns of solution to decrease transient time.[298] A solution that will attain equilibrium in 24 hours in a 3 mm column height will reach equilibrium in less than 3 h in a 1 mm column height.[299] The use of short columns simplifies the computation of molecular weight and is very useful.

TABLE 5.8 Data for Calculation of Molecular Weight by Meniscus Depletion Sedimentation Equilibrium

Reading No.	x Scale Reading (cm)	r (cm)	r^2	y Scale Reading, y(r) (cm)	$y(r) - y_0$ (cm)	$\log_{10} [y(r) - y_0] + 3$
1 (meniscus)	13.792			3.4192 ⎤		
2	13.842			3.4188		
3	13.892			3.4193 ⎬ Average (y_0) = 3.4192		
4	13.942			3.4193		
5	13.992			3.4193 ⎦		
6	14.042			3.4197		
7	14.092			3.4197		
8	14.142			3.4203		
9	14.192			3.4207		
10	14.242			3.4217		
11	14.292			3.4244		
12	14.312	6.848	46.892	3.4250	0.0058	0.763
13	14.332	6.857	47.024	3.4261	0.0069	0.839
14	14.352	6.867	47.156	3.4280	0.0088	0.944
15	14.372	6.877	47.286	3.4310	0.0118	1.072
16	14.392	6.886	47.418	3.4351	0.0159	1.201
17	14.412	6.896	47.551	3.4376	0.0184	1.265
18	14.432	6.905	47.683	3.4439	0.0247	1.393
19	14.452	6.915	47.815	3.4499	0.0307	1.487
20	14.472	6.924	47.947	3.4556	0.0364	1.561
21	14.492	6.934	48.080	3.4656	0.0464	1.667
22	14.512	6.944	48.212	3.4754	0.0562	1.750
23	14.532	6.953	48.346	3.4881	0.0689	1.838
24	14.552	6.963	48.479	3.5042	0.0850	1.929
25	14.572	6.972	48.612	3.5216	0.1024	2.010
26	14.592	6.982	48.746	3.5466	0.1274	2.105
27	14.612	6.991	48.880	3.5740	0.1548	2.190
28	14.632	7.001	49.013	3.6100	0.1908	2.281
29	14.653	7.011	49.147	3.6543	0.2351	2.371
30 (bottom)	14.697	7.032	49.443			

From Chervenka, C.H., *A Manual of Methods for the Analytical Ultracentrifuge*, Beckman Instruments, Palo Alto, CA, 1969. With permission.

The disadvantages of this technique include the following: a separate synthetic boundary run is required and conservation of mass is essential. The precision of the method is about 3%.

The data required from the equilibrium run include the concentration gradient at the midpoint of the sample column, which is obtained directly from schlieren patterns. With interference optics, the usable concentration range is extended to lower values but evaluation of results requires the measurement of the slope of the interference fringes at the midpoint of the sample column. However, one can assume that the concentration at that point is the same as the initial solution.

A 12 mm double sector cell is used with the schlieren optical system. In the right-hand sector of the cell is placed 0.01 ml of FC-43 fluorocarbon oil, followed by 0.03 ml of sample solution. In the left-hand sector is placed 0.05 ml of dialyzed solvent.

An equilibrium plate is aligned in the comparator. The radial midpoint of the image of the sample fluid column is located as half way between the upper sample-air meniscus and the lower sample-oil meniscus. The radial distance, r, is determined to this point. At this point the vertical distance between the center of the sample pattern and the center of the baseline is measured in centimeters, to give (dc/dr)mid. The synthetic boundary cell plate is placed in the comparator and a frame is selected with the same plate angle as the equilibrium photo used above. The area under the schlieren peak is measured in square centimeters to give c_0. The values of r, (dc/dr)mid, and c_0 are substituted with values for the

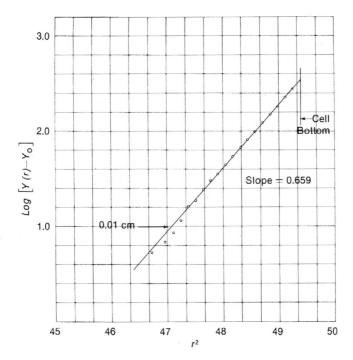

FIGURE 5.24   Log {y(r) − y$_o$} vs. r^2 plot used for the analysis of meniscus depletion sedimentation equilibrium data. (Courtesy: Beckman Instruments, Palo Alto, CA.)

partial specific volume of the solute and the density of the solution into Equation 5.34 to determine the apparent average molecular weight of the solute.

$$M = \frac{RT}{\left(1-\bar{v}\rho\right)\omega^2 \, r_{mid}c_o} \frac{1}{} \left(\frac{dc}{dr}\right)_{mid} \tag{5.34}$$

Correia and Yphantis[296] described a rapid and moderately precise short column procedure to estimate molecular weights by sedimentation equilibrium. The Beckman Model E ultracentrifuge was equipped with Rayleigh interference optics and a TRW model 83AR pulsed argon ion laser (wavelength = 5145 Å) as the interferometric light source. Interferograms were measured automatically using a Gaertner micro-comparator stage interfaced to a PDP-8L minicomputer.

Meniscus depletion experiments were performed. All experiments below 1000 rev/min were done using an An-J rotor to minimize convection. Synthetic boundary experiments were carried out in double-channel capillary centerpieces. Equilibrium times were estimated and checked empirically by comparing photographs from different times. Fringe phases and the magnitudes of the associated transforms were obtained using a modified Walsh analysis of the digitized light intensities across the interference envelope at each radial position.[300] The optimum parameters used for the Walsh analyses (the number of points per fringe, the order of the transform), the number of flashes of the laser, the delay times of the pulses, the number of fringes scanned, and the region of the fringe envelope scanned were investigated separately to maximize data precision. Data files, stored on magnetic tape, were transferred from the PDP-8L plate reading system to an IBM/370 and analyzed.

Meniscus depletion experiments were analyzed by nonlinear least-squares fitting to appropriate models using a program termed NONLIN.[296] The automated plate reading system can measure fringe displacements at 2 µm intervals, providing several hundred data points across a 0.75 mm solution column. Such

TABLE 5.9   Short Column Experimental Conditions for Some Viruses
From Which the Molecular Weight Was Calculated

Virus	Column Height (mm)	Rotor Speed (rpm)	M (millions)
Turnip yellow	1.17	1,180	5
Brome mosaic	0.9	1,156	5.3
Southern bean mosaic	1.27	1,495	6.0
Tomato bushy stunt	0.98	1,240	9.7

Note: An AnJ rotor (Beckman Instruments, Palo Alto, CA) of weight 22 lbs was used; equilibrium was achieved in 6–8 hours. The Rayleigh interference optical system was employed in a Model E Beckman Analytical Ultracentrifuge. Data from Reference 301.

sampling density is not practical with manual data acquisition. This short column approach provided a powerful and relatively precise means of estimating z-average molecular weight. A wide range of the effective reduced molecular weights, σ, defined as $M_i(1 - vp)\omega^2/RT$ can be accurately estimated. The precision of estimation of weight average values of sigma can be 1% or better.

Table 5.9 demonstrates run conditions for some viruses analyzed by the short column equilibrium method.[301]

5.1.9.5   Approach to Equilibrium Method

Approximately 50 years ago, Archibald[302] reported a method exploiting the fact that conditions for sedimentation equilibrium are achieved at the top and bottom of the cell at all times during the run. The net flow of solute at these positions is zero. From a measurement of the concentration and concentration gradient at either the upper meniscus or the bottom of the fluid column in the cell, the molecular weight of the solute can be calculated. In such a procedure the advantage is that only very short instrument operating times are required for a measurement of molecular weight. One major disadvantage is that the method depends directly upon a single measurement at the end of the fluid column. The ends represent the regions of greatest uncertainty. A synthetic boundary run measurement is required and computations are lengthy.

Measurements at the top of the cell are most useful, generally, because possible aggregate material is avoided and the concentration gradient can be measured with less error.

To calculate results, a photo from the analytical run and another from the synthetic boundary run are selected. The synthetic boundary plate is aligned in the comparator and the area under the schlieren peak is measured in square centimeters. This is c_o.

The analytical plate is set in the comparator and the radial position of the upper meniscus is measured as r_m. The vertical distance is measured in centimeters between the sample schlieren pattern and the baseline, $\partial c/\partial r$, in the center of each increment of radial distance of 0.02 cm on the comparator scale. The data are recorded as shown in Table 5.10. The first reading of $\partial c/\partial r$ is at 0.01 cm from the meniscus and subsequent readings are at 0.02 cm intervals. The schlieren pattern does not extend into the meniscus image, so the first few points will be unreadable. These readings are estimated by making a plot of the readable values of $\partial c/\partial r$ vs. r, and drawing a smooth line through the points, extending it through the meniscus position on the graph. The values of $\partial c/\partial r$ of the previously unreadable points may now be determined from the graph, including a value at the optimum meniscus position. Complete the mathematical operations indicated in Table 5.10 and sum the values of $r^2(\partial c/\partial r)$, excluding the meniscus reading from the sum to obtain:

$$\sum\nolimits_{r_m}^{r_p} r^2\left(\frac{\partial c}{\partial r}\right)$$

TABLE 5.10 Sample Data for Calculation of Molecular Weight by Approach to Equilibrium

Reading No.	x Scale Reading (cm)	r (cm)	r²	∂c/∂r (cm)	r² (∂c/∂r)
1 (meniscus)	13.975	6.687	44.716	0.2088*	—
2	13.985	6.691	44.769	0.1980*	8.864
3	14.005	6.701	44.903	0.1760*	7.903
4	14.025	6.710	45.024	0.1556	7.006
5	14.045	6.720	45.158	0.1338	6.042
6	14.065	6.730	45.293	0.1111	5.032
7	14.085	6.739	45.414	0.0897	4.074
8	14.105	6.749	45.549	0.0673	3.065
9	14.125	6.758	45.671	0.0520	2.375
10	14.145	6.768	45.806	0.0351	1.608
11	14.165	6.777	45.928	0.0231	1.061
12	14.185	6.787	46.063	0.0127	0.585
13	14.205	6.797	46.199	0.0051	0.236
14	14.225	6.806	46.322	0.0012	0.056
15 (plateau)	14.245	6.816	46.458	0	0

* Estimated from a plot of $\partial c/\partial r$ vs. r.
From Chervenka, C.H., *A Manual of Methods for the Analytical Ultracentrifuge*, Beckman Instruments, Palo Alto, CA, 1969. With permission.

This value is substituted into Equation 5.35 to compute the concentration at the meniscus, c_m. Use 0.02 for Δx, and the magnification factor for the plate for F. The value obtained for c_m, along with appropriate values for $(\partial c/\partial r)_m$, r_m, and the values for ω T, $\bar{v}$, and ρ in Equation 5.36 are used to determine the apparent molecular weight of the solute at the meniscus.

If the schlieren pattern permits, the above calculation can be repeated for the bottom of the cell. The calculation for the concentration of solute at the bottom meniscus is carried out analogous to that used for the upper meniscus, except that values of r_b, c_b, and $(\partial c/\partial r)_b$ are used in Equations 5.35 and 5.36 instead of the meniscus values. Also, in Equation 5.35, the integral on the right is added to the initial concentration instead of subtracted from it.

$$c_m = c_o - \frac{1}{r_m^2}\int_{r_m}^{r_p} r^2\left(\frac{\partial c}{\partial r}\right)dr$$

$$= c_o - \frac{1}{r_m^2}\left(\frac{\Delta x}{F}\right)\sum_{r_m}^{r_p} r^2\left(\frac{\partial c}{\partial r}\right)$$

(5.35)

$$M = \frac{RT}{(1-\bar{v}\rho)\omega^2}\frac{1}{r_m c_m}\left(\frac{\partial c}{\partial r}\right)_m$$

(5.36)

An example of the calculation of molecular weight by approach to equilibrium[234] is as follows:

$(1 - \bar{v}p) = 0.308$, rotor speed = 20,000 rpm, temperature = 19.2°C;

initial concentration = 0.0713 cm², r = 6.687 cm.

$(\partial c/\partial r)_m = 0.2088$ cm, F = 2.095, Δx = 0.020.

Summing the last column in Table 5.10 = 47.907
From Equation 5.35,

$$c_m = 0.0713 - 1/44.716 \ (0.020/2.095) \ 47.907 = 0.0611$$

Using Equation 5.36

$$RT/(1 - \bar{v}p) \ \omega^2 = (8.313 \times 10^7) \ 292.4/0.308 \ (4.387 \times 10^6) = 1.795 \times 10^4$$

$$M = (1.795 \times 10^4) \ 0.2088/6.687 \times 0.0611 = 9.173 \times 10^3$$

The Beckman Optima series of analytical ultracentrifuges raise sedimentation studies to higher levels of analysis. In this connection the Archibald procedure may find renewed application with absorbance optics and appropriate software.[227]

The molecular weight of many viruses were obtained by analytical sedimentation. Table 5.11 presents some representative viruses analyzed by this procedure.

TABLE 5.11   Molecular Weight Values for Viruses from Analytical Sedimentation Experiments

Specific Virus	M (millions)	$s_{20,w}$	$D_{20,w}$*	$\bar{v}$
Animal				
Avian myeloblastosis[303]	256	619	0.268	0.78
Eastern equine encephalitis[304]	58	240		
Encephalomyocarditis[305]	8.5	162	1.44	0.68
Feline leukemia[306]	550	880	0.255	
Foot-and-mouth disease[307]	8.3			
Infectious pancreatic necrosis[308]	55	435	0.67	0.706
Influenza, PR8[309]	360			
Kilham rat[310]	6.6	122		
Mouse encephalomyelitis[311]	11.5	152		
Murine mammary[306]	317	595	0.299	0.847
Polio[312]				
Mahoney	6.8	160		0.637
MEF 1	6.8	158		0.641
Saukett	6.4	157		0.617
Polyoma[313]	25	240		
Rauscher murine leukemia[314]	390	625	0.296	
Reo[315]	70	630		
Rhino[316]	7.1	158	1.71	0.682
Rhino 1A[317]	8.4			
Rous sarcoma[303]	294	739	0.29	0.79
Sindbis[318]	50	300		
Syrian hamster oncogenic papovavirus[319]	27.5	233		
Vesicular stomatitis[320]	290	610	0.233	0.78
Bacterial				
f2[318]	4	80		
fd[311]	11.3	40		0.71
fr[311]	4.1	79	1.45	0.69
Hy ϕ30[321]	55.4	492		
IF1[322]	25	45		
Lambda[323]	57	416		
MS2[324]	5.3	81.5		
N4[325]	83	615		
ϕ 6[326]	99			
$\phi \times 174$[327]	6.2	114		
P11-M15[328]	66.7			
PB-2[329]	67.5	60		
PL25[330]	54.3	485	0.68	

TABLE 5.11   Molecular Weight Values for Viruses from Analytical Sedimentation Experiments

Specific Virus	M (millions)	$s_{20,w}$	$D_{20,w}$*	$\bar{v}$
PM2[282]	47.9	294	0.65	0.771
QB[282]	4.55	88.4	1.423	0.668
σ A[322]	17	46		0.769
R17[277]	3.83	c 75	1.47	0.673
T2[331]	220	1066		
T3[332]	49	476		
T4[266]	192.5	890	0.295	0.618
T5[266]	109.2	615	0.397	0.658
T6[333]	145	1050		
T7[266]	50.4	453	0.603	0.639
TP-84[334]	50	436		
Insect				
Densovirus of *Galleria mellonella* (DNV-1)[336]	5.7	117		0.674
Drosophilia X[308]	81.2	503	0.610	0.752
Plant				
Bean pod mottle[337]	5.7	112		
Broad bean mottle[338]	5.2	84.8		0.717
Brome mosaic[339]	4.6	86.2		0.708
Carnation mottle[340]	7.7	118	1.24	0.700
Cowpea chlorotic[341]	4.6	88	1.46–1.54	0.695
Cucumber necrosis[342]				
Major	9.3	136		
Minor		50		
Echten ackerbohnen mosaic[311]	6.4	119		
PBCV-1[335]	1000	2300		
Potato X[318]	35	118		
Squash mosaic[343]				
Top	4.5	57	1.28	0.75
Middle	6.1	95	1.35	0.69
Bottom	6.9	118	1.32	0.67
Southern bean mosaic[344]	6.6	115		0.70
Tomato bushy stunt[345]	8.9	131	1.246	0.71
Tobacco mosaic[318]	40	190		
Tobacco necrosis satellite[346]	1.9	50		
Tobacco rattle[347]				
l particles	48–50	296–306		
s particles	11–29	155–245		
Tobacco ringspot[348]		116		
Turnip yellow mosaic[293,349]	5.4	106	0.67	

* $D_{20,W} = 1 \times 10^{-7}$ cm^2/s.

5.2   Density Gradient Centrifugation*

Frequently the preparative ultracentrifuge is used to obtain values of density, sedimentation coefficients, and the molecular weights of bioparticles. In tubes set at an angle and with parallel walls, convection would disturb the smooth migration of particles that is required for analysis or complete separation. Employing gradients, however, preparative ultracentrifugation may be useful for experiments formerly limited to the analytical instrument.[350,351]

---

* Much of this discussion is contained in *An Introduction to Density Gradient Centrifugation — A Technical Review*. 1960. Published by Spinto Division, Beckman Instruments, Inc., Palo Alto, California.

Density gradient experiments are generally classified into three categories according to the way the gradient is used. The first category is called *stabilized moving-boundary centrifugation*. It is analogous to classical ultracentrifugation experiments in which the sedimentation coefficient is determined. In the preparative ultracentrifuge the sample is separated into fractions, which are analyzed to determine the position of the boundary, from which the sedimentation coefficient is calculated. The gradient used is quite shallow, and its function is to stabilize against convection. The solution containing the particles to be measured is distributed throughout the gradient column at the start of the experiment.

The second form of density gradient experiment is referred to as *zone centrifugation*. Particles having different sedimentation rates are completely separated. In zone centrifugation a solution containing particles of different characteristics is layered on top of the gradient column before centrifugation. Since layering additional material at the top of the gradient column creates an unstable negative gradient just below the zone of particles, the density gradient for zone separation must be quite steep. The steep gradient serves to reverse the negative gradient in a very short distance, thereby keeping convective stirring at a minimum. Each type of particle, sedimenting at its own rate, then forms a band or zone in the fluid column. These zones of solute will then be separated from one another by distances related to their sedimentation rates. After centrifugation, each solute type can be drawn off separately for the determination of its sedimentation rate and for further analysis.

The third form of density gradient experiment is called *isopycnic (isodensity) gradient centrifugation*. This technique is based solely on the different densities of the various biosolutes. The molecular weight and partial specific volumes of the particles may also be calculated. In this type of experiment, the gradient column encompasses the entire density range of the particles, and sedimentation is continued until the particles reach positions at which the density of the surrounding liquid is equal (iso, the same) to their own. The solution containing the particles of interest is usually distributed evenly throughout the gradient column before centrifugation, or it may also be layered at the top or bottom of the column.

Density gradient centrifugation has a number of virtues: (1) it is relatively easy to perform; (2) very low, concentrations can be used; (3) the limit of sensitivity is governed only by the detectability of the sedimenting particles. Comparison of results with those obtained with the analytical ultracentrifuge, and with techniques other than centrifugation, demonstrates that density gradient centrifugation provides results that are accurate.

5.2.1   Stabilized Moving-Boundary Centrifugation

In 1943, Pickels[352] studied the use of angle head rotors and concluded that serious convective disturbances precluded an interpretation of the sedimentation properties of macromolecules. During the course of his investigation, Pickels found that convection could be virtually eliminated through the presence in the tube of a slight gradient of density caused by some light, rapidly diffusing substance such as sucrose. In a subsequent study, Friedewald and Pickels[353] studied the PR-8 and Lee strains of influenza virus by using a density gradient in an angle head rotor. The sedimentation boundaries of infective virus particles, hemagglutinin, and complement fixing antigen correlated with the boundaries observed in the analytical ultracentrifuge.

Seeking to improve the precision with which sedimentation coefficients could be measured in a preparative rotor, Kahler and Lloyd[354] began to work with a rotor containing buckets, which could pivot from vertical to horizontal during centrifugation. In this configuration, the direction of centrifugal force would always be parallel to the walls of the buckets. They concluded that such swinging bucket rotors possessed great advantages over fixed-angle rotors for measuring sedimentation rates. However, the problem of convection still remained as a result of radial sedimentation in a parallel-walled bucket. The use of density gradients was necessary if boundaries were to be maintained. By using the swinging bucket rotor along with a stabilizing density gradient, the value of the sedimentation coefficient of polystyrene latex was in excellent agreement with that obtained with the analytical ultracentrifuge.

Hogeboom and Kuff[355] then used the swinging bucket rotor to make a very thorough study of a variety of proteins. They were able to determine sedimentation coefficients over a range of 4 to 2000 S with a high degree of accuracy. By use of a shallow sucrose gradient, they were able to determine the sedimentation coefficients of materials even in very dilute solutions.

5.2.2 Zone Centrifugation

Complete separation of several components in a mixture is not achieved with the moving boundary method. In a study by deDuve and Berthet in 1954,[356] it was observed that when the bottom of the tube is twice as far as the meniscus from the axis, the sedimented pellet containing all of the fastest moving component will be contaminated with 40% of a component having 1/3 its speed, and 13% of a component having 1/10 its speed.

The need for another kind of sedimentation technique capable of separating macromolecules with greater resolution was apparent. Brakke[357] devised a technique in which a solution of particles is layered on top of a density gradient column prepared from some solvent and solute in which the particles are soluble. The centrifuge tubes are then placed in a swinging bucket rotor and centrifuged at high speed. Although the addition of this layer creates a negative gradient at the top of the column, the steep positive density gradient supports the layer of particles, preventing the sedimentation of droplets of liquid and ensuring that the particles will migrate at a steady rate.

The particles separate into zones as they migrate through the column under the influence of centrifugal force. Each zone will consist of one type of molecule, moving through the gradient at its characteristic sedimentation velocity. If the centrifuge is stopped while the molecules are still sedimenting, the molecules will then be separated according to the differences in their sedimentation rates. This separation technique exploits the effect upon sedimentation rate of differences in the size, shapes, and densities of the macromolecules.

The technique of zonal centrifugation has been evaluated.[358] The advantages are (1) only small amounts of material are needed; (2) there is little dependence of s on concentration; (3) the sample sediments free of interfering substances. The disadvantages noted are (1) inability to measure an absolute s value; (2) poorer resolution than that in the analytical ultracentrifuge.

In zone centrifugation the rate of migration of the center of mass $\langle r \rangle$ is the quantity that is exactly related to the sedimentation velocity of the macromolecules, as shown by Schumaker and Rosenbloom.[359]

To obtain the sedimentation coefficient of viruses, Brakke[360] devised a procedure using as a standard a virus whose sedimentation coefficient is known. The standard virus serves as a marker in the same centrifuge tube as the unknown virus, or a different tube in the same rotor. A plot is constructed of $\langle r \rangle$, the center of mass vs. $\int \omega^2 \, dt$ for both viruses (Figure 5.25). Smooth curves are drawn through the points. The values $\int \omega^2 \, dt$ are read from the curves for both virus zones at the same $\langle r \rangle$ locations. The ratio of these values, $\int \omega^2 \, dt$, is equal to the inverse ratio of the sedimentation coefficient multiplied by a small term involving the partial specific volumes. To eliminate this small term, the ratio of the values of $\int \omega^2 \, dt$ is plotted as a function of $(\rho + \rho_{20,w})/2$, and extrapolated to $\rho_{20,w}$. The extrapolated value then will be exactly the ratio of sedimentation coefficients (Figure 5.26).

Freifelder[358] cautioned on the use of adding a single standard to a zonal run. He advocates that with two or preferably three markers, one can empirically relate all the distances sedimented. He cites two principal limitations in the determinations of a relative s value in zonal centrifugation: (1) in measuring relative distance one must know where to choose the sedimentation origin. This may not be done with certainty because of the finite thickness of the original sample layer and also because of the accumulation of material at the sample-gradient interface. In practice this difficulty may introduce an error of about 2% to 3% in the s value. (2) A more serious problem arises from the finite size of the fraction taken. This limits the accuracy of identification of the fraction containing a "peak" and probably introduces an error of +5%. The usual reason for measuring is to calculate the molecular weight. He notes that the s

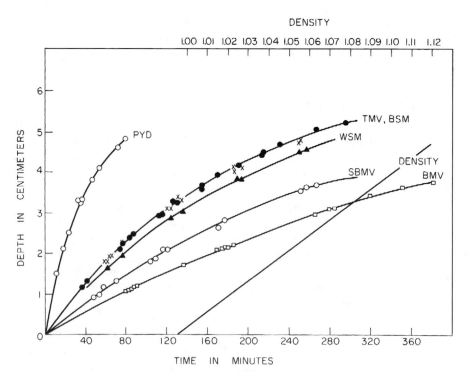

FIGURE 5.25   Sedimentation of viruses through a gradient column prepared from 3, 7, 7, 7, and 4 ml of solutions of 0, 100, 200, 300, and 400 mg of sucrose/ml, respectively, in a 1 × 3 in. centrifuge tube. Centrifugation was at 24,000 rpm for various periods, at 5–6°C, in the SW 25.1 rotor of the Spinco (Beckman) Model L Preparative ultracentrifuge. Two milliliters of virus solution, containing approximately 0.02–0.2 mg of virus, were floated on each tube. One tube with tobacco mosaic virus (TMV) as a standard was included in each tube. The depth of TMV in this tube was used to estimate the effective time of centrifugation from a preliminary curve constructed from results of the first run. The effective time of centrifugation is the time that would have been required for the virus to sediment to the observed depth if conditions had been the same as the first run. Except for the first run, all points are plotted at the effective time of centrifugation rather than the actual time. Depths were measured from the meniscus to the top of the light scattering zone. PYD, potato yellow dwarf virus; TMV, tobacco mosaic virus; BSV, barley stripe mosaic virus; WSM, wheat streak mosaic virus; SBMV, southern bean mosaic virus; BMV, brome mosaic virus. (From Brakke, M.K., *Virology*, 6, 96, 1958. With permission.)

measured by zonal centrifugation is not the s used in theoretical equations. The s measured by zonal centrifugation is always less than the s value from the analytical ultracentrifuge. Therefore, it is essential that in zonal centrifugation one uses empirical relations derived from measurements in which three or four substances of known molecular weight be cosedimented with the particle of unknown molecular weight.

5.2.3   Isopycnic Gradient Centrifugation

This method of separating particles by densities makes use of a gradient within the tube extending over the whole range of densities of the macromolecules to be analyzed. The particles layered at the top will not reach the bottom of the tube even after prolonged centrifugation. Rather, they will attain equilibrium positions at levels in the tube corresponding to their own densities. This method of separating macromolecules by densities has been termed isopycnic gradient centrifugation by Anderson.[361] Earlier, the method was used by Linderstrom-Lang and Lanz[362] for determining the densities of microdrops of liquid. It is still a popular method for determining the densities of protein and virus solutions and for the calculation of partial specific volumes.

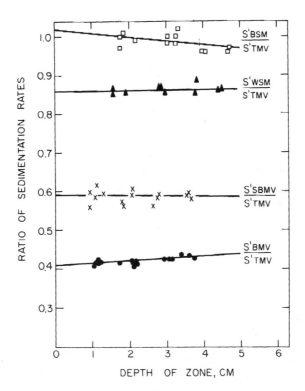

FIGURE 5.26 Ratio of the sedimentation rates (S′) of various viruses to the sedimentation rate of tobacco mosaic virus plotted against the depth of the zone of the first virus. Experimental conditions and symbols are the same as in Figure 5.25. (From Brakke, M.K., *Virology*, 6, 96, 1958. With permission.)

The macromolecules are centrifuged through a solvent containing a steep gradient of salt, sucrose, or other low-molecular-weight solute. Eventually the large molecules approach that point in the centrifuge tube where the solution densities will be equal to the buoyant density of the particles. At this point, the term $(1 - \bar{v}p)$ in Equation 5.25 will become zero, and sedimentation will stop. At the conclusion of the run, the macrosolutes will be found in a narrow zone centered at this isopycnic (or isodensity) location in the centrifuge tube.

Historically, this technique has been used primarily for the analysis of nucleic acids. Meselson and Stahl[363] used the method to detect small density differences brought about by replacement of ^{14}N with ^{15}N. The method has led to significant understanding of the semiconservative replication of DNA.

Equations for calculating molecular weights from measurements of the positions and shapes of the bands, as well as the properties of the material forming the gradient, were derived by Meselson et al.[364] However, Baldwin[365] and Baldwin and Van Holde[366] have shown that heterogeneity in density among the molecules can cause substantial errors in molecular weights calculated by this method. Also, preferential interactions between the molecules and either the solvent or the gradient material will also cause uncertainty in the values obtained.

Ift and Vinograd[367] noted that in density gradient work, the molecular weight and specific volume of the macromolecule, as determined from the position of the band, are not those of the anhydrous species because of substantial interaction with the gradient-forming solute. Because the buoyant density is determined in a solvent that is very different from dilute aqueous solutions, its inverse should not be used to approximate the partial specific volume that is appropriate for dilute aqueous solutions.

5.2.4 Use of Zonal Rotors[368]

Zonal rotors were discussed in Chapter 3. The use of zonal rotors in density gradient experiments allows very much larger volumes to be run over what can be analyzed with swinging bucket or angle head rotors.

From a knowledge of run time, rotor speed, and gradient concentration it is possible to estimate representative values for sedimentation coefficients of bands of sample components in rate zonal runs. Done in tubes, these calculations are complicated by the presence of the gradient through which the particles sediment, and are generally uncertain because the effective particle density is not known, or it changes with the concentration of gradient material.

In zonal rotors, the radial position of the middle of the starting zone, r_s, is determined from a knowledge of the shape and volume relationship of the rotor and from the volumes of sample and overlay used. The average radial position which each fraction, collected at the end of the experiment, occupied in the rotor, r_j, and the radial position of the sample band, r_p, are determined from the effluent volumes. The concentration of gradient material in each fraction is measured and its density, ρ_j, and viscosity, η_j, for the experimental temperature are determined from appropriate tables.

Sedimentations coefficients are calculated using:[369]

$$s_{20,w} = \frac{\rho_p - \rho_{20,w}}{\omega^2 t\, \eta_{20,w}} \sum_{r_p}^{r_s} \frac{\eta_i\, \Delta r_i}{\left(\rho_p - \rho_i\right) r_i} \qquad (5.37)$$

where ρ_p is the density of the sample particles, $\rho_{20,w}$ is the density of water at 20°C (0.998 g/cm^3), $\eta_{20,w}$ is the viscosity of water at 20°C (0.010 poise), and Δr_i is the radial width of each affluent fraction (the radial distance through which the particles sedimented in each fraction). The quantity $\omega^2 t$ is the product of the square of the angular velocity of the rotor in radians per second and the run time in seconds. The relation between ω and revolutions per minute is

$$\omega = \frac{2\pi\, \text{rpm}}{60} = 0.1047\, \text{rpm} \qquad (5.38)$$

To compensate for the changing force during acceleration and deceleration, the run time is measured between the time when the rotor reaches two thirds of the operating speed during acceleration and the time when it returns to this speed during deceleration. In order to be exact, the quantity $\omega^2 t$ should be replaced with the time integral of the square of the rotor velocity, which can be determined directly by the use of the electronic $\omega^2 t$ accessory. Equation 5.37 gives values of the sedimentation coefficient in seconds; to convert to Svedberg units, divide by 10^{-13} s.

Preparative ultracentrifuges have an important role in density gradient sedimentation. However, analytical instruments are also used. Vinograd et al.[370] demonstrated that it is possible to generate a stabilizing density gradient in an analytical ultracentrifuge cell with the aid of a band-forming cell. In a band-forming cell, a narrow zone of solution containing the biosolute is layered over a solution containing an auxiliary solute such as cesium chloride. The density of the salt solution prevents the gross convection of the layer of macromolecule solution to the cell bottom. As sedimentation proceeds, macromolecules from the layer sediment into the salt solution. Diffusion of solvent from the layer and some degree of sedimentation of the salt combine to maintain a self-generating density gradient that stabilizes the sedimenting zones. Each zone can then be observed as a sedimenting bell-shaped profile of absorbance, utilizing the scanner absorption optics. This method is well suited to study DNA, with its high absorbance coefficient. However, since the density and viscosity of the supporting density gradient change slowly as the density-generating solute redistributes in the gravitational field, it is difficult to obtain absolute sedimentation coefficients by this method.[371]

Density gradient centrifugation has been commonly employed in the study of various particles associated with viral infection, e.g., the complete virus particle, the A or 135s particles, or particles called provirions. Sedimentation coefficients were determined for the particles present in the density gradient zones. Isolation of the fractions permitted studies to determine the precise role of the particles in the infection process. Table 5.12 lists some representative viral systems studied by density gradient centrifugation.

TABLE 5.12 Some Viruses and Associated Particles Studied by Sucrose Density Gradient Centrifugation

Virus/Associated Particle	Sedimentation Coefficient	Ref.
Adeno 2, 3, 5, 7, and 21	770	372
Coxsackie B:A particles	80–160	373
Epstein-Barr:nuclear antigens	13 and 34	374
Flockhouse:mature paticle, provirion	137	375
Flockhouse:recombinant (nodavirus)	130	376
Hemorrhagic disease-rabbit	100,136,175	377
Polio 1	135	378
Polio 1	150	379
Rhino 14-human	90–149	380
Rhino 14-human	150,125,80	381

5.3 Supplementary Procedures for Obtaining Molecular Weight from Values of Density, the Partial Specific Volume, Viscosity, and the Frictional Coefficient

In addition to recording the concentration distribution of the virus, other quantities are needed to obtain the molecular weight value:[382] the partial specific volume of the virus and the density of the solvent. In order to correct for the effect of different solvents and temperatures on sedimentation behavior, the viscosity of the solvent and its temperature dependence are required. In many runs these values are obtained from tables. However, for accurate results, the quantities should be measured.[383,384]

5.3.1 Density

The fluid density, ρ, of solutions in sedimentation analyses is required for calculations of molecular weight. For 1% accuracy in such calculations, the sample density must be known within 0.3%. Commonly, densities of dilute solutions are measured pycnometrically with 10 to 25 ml volumes.[385,386] The addition of the virus solute to the buffer or salt solution does not change the density significantly, so for approximate measurements, the density of the solvent alone may be determined, thus avoiding the requirement for relatively large quantities of sample. For more accurate determinations the density of the solvent can be determined, and then a computation made to obtain the density of the solution. Another technique is to use volumetric pipettes of 0.5 to 2.0 ml volume. These are filled with solution at room temperature, weighed to the nearest milligram, and the temperature recorded. From the net weight of the solution and the volume of the pipette, the density at the temperature is calculated.

When using solvents of high concentration, densities are best determined indirectly from a knowledge of the concentration. The concentration is determined refractometrically and then referred to appropriate tables to obtain density. Some investigators have supplied equations to aid in obtaining density from refractometric measurements.[387]

5.3.2 Partial Specific Volume, $\bar{v}$

This quantity, closely related to density, is defined thermodynamically as the limit in the change in volume of solution per change in weight of the dry constituent as the changes approach zero.[243] Operationally, $\bar{v}$ is the volume of solvent, usually water or dilute salt solution, displaced by the addition of virus. The unit for $\bar{v}$ is ml/g, and $\bar{v}$ is approximately the reciprocal of the dry density of the particle. An error of 1% in $\bar{v}$ leads to an error of approximately 3% in molecular weight.[227]

Some methods employed for the determination of $\bar{v}$ are presented. In one method, the chemical composition in terms of protein and nucleic acid are considered. The $\bar{v}$ value for proteins is in the range 0.70 to 0.75, and for nucleic acid, 0.45 to 0.55. Values for the protein component may be obtained from a knowledge of the amino acid composition sequence.[388,389] The percentage of the amino acids multiplied by the $\bar{v}$ of each amino acid, obtained from tables, and summing this product for all amino acids in the protein gives values for $\bar{v}$ of the protein component. The relative weighted $\bar{v}$ values for the protein and nucleic acid components are added to give the $\bar{v}$ value for the virus.

It has been pointed out that $\bar{v}$ cannot always be predicted simply by addition of values of the $\bar{v}$ of components. The use of a calculated rather than a measured value for $\bar{v}$ may introduce much uncertainly when $\bar{v}$ is calculated from the amino acid composition and the percentage of DNA. For coliphage T5, the value for $\bar{v}$ is not that which would be obtained from such calculations.[390,391]

The partial specific volume of the virus may be calculated from solution densities measured by means of the method of Linderstrom-Lang and Lanz.[362] A density gradient thermostated at constant temperature is formed by any of the various density gradient procedures, employing chlorobenzene and 1-chlorobutane, saturated with buffer. The appropriate range of densities, as determined by the approximate concentration of the solution under study, is chosen, and standard salt solutions, e.g., KCl, are prepared with densities covering this range. The two organic liquids are mixed in the appropriate proportions to give two solutions, one in which a droplet of the most dense solution just floats and one in which that of the smallest density just sinks. These two solutions are then used to make the gradient. Droplets of the standards and the sample are introduced into the volume with the aid of a capillary tube. The droplet will come to rest at a level corresponding to its density, which is calculated by interpolation between the positions of standards. After a short equilibration period (15 minutes) their distances from a fixed point at the top of the column is determined with a cathetometer. The concentration of the virus solutions is determined by dry weight. The samples are brought to a constant weight by heating in a vacuum oven at a selected temperature setting.

Another method for determining $\bar{v}$ is by pycnometric means. Here, the amount of solvent displaced by a known weight of virus is ascertained. The temperature of the pycnometer is kept to a known value, to within $\pm 0.05°C$. The virus solution and reference solvent are placed in a relatively large vessel of water, and allowed to equilibrate. Careful stirring of the water is done to avoid heating. The volume of the pycnometer need not be known as long as the total amount of virus contained in it is known. The pycnometer is first filled with the pure solvent, which is allowed to drain and equilibrate. The level is then accurately adjusted by means of a fine capillary tube, stoppered, dried, and weighed. Most of the solvent is removed with a syringe, and then a known weight of virus dissolved in and previously equilibrated with solvent is added. Pure solvent is added to the mark and the level is adjusted as before. This time the weight will be greater than the weight of the solvent required to fill the difference in volume between that of the virus and *of the same weight of solvent.* This difference is $\bar{v}$.

In early sedimentation studies, Svedberg and Eriksson-Quensel[392] found that $p°$ for hemocyanin corresponded closely to the reciprocal of the partial specific volume. The parameter $p°$ is the density of the solution in which the solute particles neither sediment nor float; $s = 0$. A number of investigators have used this relationship for the determination of the partial specific volume of viruses.

As noted, the partial specific volume, $\bar{v}$, is the volume of water displaced by 1 g of virus in a dilute aqueous salt solution. It enters into the Svedberg equation as the factor $(1 - \bar{v}p)$, where p is the density of the solution.

The determination of the density of a solution in which the sedimentation coefficient is equal to zero gives a measure of $\bar{v}$. To obtain $\bar{v}$, the sedimentation rate of the virus is measured in a series of increasing solution density. The data are plotted and extrapolated to the value of p, corresponding to zero sedimentation rate. This value for the density of the solution is equal to $1/\bar{v}$. In these experiments, mixtures of D_2O and water in various preparations can be used. A drawback of the method is that extrapolation of the data is necessary to reduce the sedimentation rate of the biosolute to zero. Undiluted D_2O is not sufficiently dense to obtain a zero sedimentation rate experimentally.

Sedimentation rates are carried out in buffer containing 50%, 70%, and 100% D_2O. Apparent sedimentation coefficients (S_{app}) are calculated from the radial displacement of the peak maxima. The viscosities and densities of the sedimentation solvents are measured. Values of ($S_{app} \eta/\eta_o$) are plotted against p, the solvent density, and $\bar{v}$ is obtained as the reciprocal of the intercept at $S_{app} \eta/\eta_o$ and zero by the method of least squares.

Investigators have noted that for a monodisperse multicomponent macromolecular solution it is important to consider interactions between biosolute and solvent.[282,389] The net effect of these interactions on the apparent molecular weight is termed preferential interactions. The effect of these interactions on molecular weight determinations may be eliminated by defining an apparent specific volume ϕ' related to the apparent density increment at constant chemical potential of all diffusible components ($\Delta\rho/c)_\mu$, as discussed by Cassasa and Eisenberg:[383]

$$\left(\Delta\rho/c\right)_\mu = 1 - \phi' \rho_o \qquad (5.39)$$

where c is the concentration of macromolecule and Δp is the difference between the density of the macromolecular solution taken undiluted after dialysis, ρ_s, and that of the dialysate, ρ_o. It is assumed that the value of ϕ' obtained is equal within experimental error to its value at infinite dilution. Densities may be determined pycnometrically. Concentration values may be determined spectrophotometrically from optical densities and extinction coefficients.

Edelstein and Schachman[393] devised an alternative method for estimation of both $\bar{v}$ and molecular weight from sedimentation equilibrium experiments in the analytical ultracentrifuge. Two sets of data are obtained: one set of data of c vs. r is obtained with water as a solvent. The second set is obtained with a mixture of D_2O/H_2O of known density to yield $\bar{v}$, which is then used to calculate molecular weight.

5.3.3 Viscosity[394]

Viscosity is the resistance that a fluid exhibits to the flow of one layer over another. The coefficient of viscosity η of a liquid can be determined directly in poises by passing a liquid through a tube of small diameter and making use of the following formula:

$$\eta = \frac{p\pi r^4}{8vl} t \qquad (5.40)$$

where t is the time required for v ml of liquid to flow through a capillary tube of length l and radius r under an applied pressure P. The quantitative measurement of absolute viscosity by this method is difficult. In practice, indirect measurements are usually made in which the viscosity of a liquid is determined relative to another liquid whose viscosity is already known. The determination of physical chemical constants by making relative measurements of a standard substance is a common procedure.

In a viscometer the pressure that causes flow through the capillary depends upon the difference in height h of liquid levels in the two tubes, the density of liquid d_1, and the acceleration of gravity g. The equation may be rearranged to

$$\eta_1 = \left(\frac{hg\pi r^4}{8lv}\right) d_1 t_1 \qquad (5.41)$$

If the same volume of a second liquid of viscosity η_2 and density d_2 takes t_2 seconds to flow through the same capillary,

$$\eta_2 = \left(\frac{hg\pi r^4}{8lv} \right) d_2 t_2 \tag{5.42}$$

In taking the ratio of these equations the height h cancels if the same volume of the two liquids has been used, and so do the other instrumental constants, so that

$$\eta_1 / \eta_2 = d_1 t_1 / d_2 t_2 \tag{5.43}$$

From measurements of the densities and times of flow of the two liquids it is a simple matter to calculate the viscosity η_1 of the first liquid when that of the second is known. In sedimentation analyses, water is taken as the reference liquid, and at 20°C, has a value of 0.01002 poise.

5.3.4   The Molar Frictional Coefficient[395]

The centrifugal force acting on molecules is balanced by a frictional force, (f/N) dx/dt, where f is the molar frictional constant for the molecule and dx/dt is the sedimentation velocity. The molar frictional constant is related to s, the sedimentation constant and M, the molecular weight:

$$f = M\left(1 - \bar{v}p\right) / s \tag{5.44}$$

The molar frictional constant is also related to D, the diffusion constant:

$$f = RT / D \tag{5.45}$$

where R and T have their usual significance.

When the molecular weight is known it is possible to calculate what f would be for a compact spherical and unhydrated molecule of the same mass:

$$f_o = 6\pi\eta N \left(\frac{3Mv}{4\pi N} \right)^{\frac{1}{3}} . \tag{5.46}$$

If the frictional ratio f/f_o is equal to 1.0 for a given solute, the shape of the molecules is spherical and the molecules are not hydrated to any measurable extent. If f/f_o is larger than l, the macromolecules either deviate from a spherical shape or are hydrated. The macromolecules may be asymmetrical as well as hydrated.

In theory, any method by means of which f/f_o may be determined will, by combination with s or D, permit the calculation of M. Values of viscosity data may also be related to s and M.

From specific viscosity values and sedimentation data, Lauffer, in 1938, calculated a molecular weight value for TMV.[396,397] The theoretical background for this procedure starts with Einstein's derivation of an equation relating the viscosity, η, of a dilute suspension of small rigid spheres to the volume fraction, ϕ, occupied by the spherical particles:[398]

$$\frac{\left(\eta/\eta_o\right) - 1}{\phi} = \frac{\eta_{sp}}{\phi} = 5/2 \tag{5.47}$$

where η_o is the viscosity of the solvent. The quantity $(\eta/\eta_o) - 1$ is the specific viscosity, η_{sp}, and 5/2 is the viscosity coefficient for spheres. For dilute suspensions, viscosity, in this equation, is independent of the size of the spheres.

Lauffer determined the specific viscosities of aqueous solutions of TMV and used a derived equation for the viscosity of randomly oriented rod-like particles such as TMV:

$$\frac{\eta}{\eta_o} - 1 = 2.5\phi + \frac{\phi}{16}\left(\frac{b}{a}\right)^2 \tag{5.48}$$

where b/a is the ratio of length to diameter of the particles. An equation expressing the dyssymetry factor used by Svedberg, f/f_o, as a function of the ratio of the major to the minor axes of particles as rod-like ellipsoids of revolution is

$$\frac{f_o}{f} = \frac{\left(a/b\right)^{\frac{2}{3}}}{\sqrt{1-\left(a/b\right)^2}}\log_e\frac{1+\sqrt{1-\left(a/b\right)^2}}{a/b} \tag{5.49}$$

Knowing the dyssymmetry factor, the molecular weight of the suspended particles may be calculated with the equation

$$\frac{f}{f_o} = \frac{M\left(1 - vd\right)}{6\pi\eta_o N s_{20}\left(3MV/4\pi N\right)^{\frac{1}{3}}} \tag{5.50}$$

where M is molecular weight, $\bar{v}$ is the partial specific volume of the virus particle, taken as 0.73, d is the density of the solvent, N is Avogadro's constant, and s_{20} is the sedimentation constant (uncorrected), taken as 174×10^{-13}. A value for M of 42.6 million was calculated for TMV.

It should be noted that in Lauffer's original report in 1938, TMV is referred to as tobacco mosaic virus protein. In 1938, in most cases, viruses were referred to as proteins, following Stanley's announcement in 1935 that TMV was a protein. This use of a misnomer was corrected after Bawden and Pirie's finding that TMV also contained nucleic acid. However, for a few years, virus particles continually were cited as "virus proteins." For example, sedimentation coefficients for a number of viruses were referred to as virus proteins, when in reality what was measured was the complete (nucleoprotein) virus.[224]

In viscosity studies there is difficulty in knowing the fraction of the volume, ϕ, occupied by high polymer molecules in solution. Moreover, the ratio of n_{sp} to concentration is not independent of concentration. It is necessary to utilize the intrinsic viscosity, [n], defined as $[n] = \lim_{c\to 0}(n_{sp}/c)$, where c is the concentration of the high polymer in g/100 ml. The intrinsic viscosity is obtained by plotting the ratio of specific viscosity to concentration against concentration for a series of solutions and extrapolating to zero concentration.

The intrinsic viscosity together with the corrected sedimentation coefficient may be used to calculate the molecular weight by means of the Scheraga-Mandelkern equation:

$$M^{\frac{2}{3}} = \frac{s_o[\eta]^{\frac{1}{2}}\eta_o N}{\beta\left(1 - \bar{v}\rho_o\right)} \tag{5.51}$$

where N is Avogadro's number and β is the shape parameter, the minimal value of which is 2.12×10^6 for spheres.

The molecular weight of N4 coliphage was also calculated using Equation 5.51. A value of 89.3 × 10[6] was obtained. This value was in close agreement with M values calculated from light scattering data and from sedimentation and diffusion analyses.[399]

Marvin and Hoffman-Berling[311] also made use of the Scheraga-Mandelkern equation to calculate a molecular weight value of 4.1 × 10[6] for the spherical RNA phage, fr. This value corresponded closely to that calculated from the empirical equation derived by Marvin and Hoffman-Berling for obtaining molecular weight from sedimentation and $\bar{v}$ (see next section).

5.4   Empirical Equations Relating Molecular Weight and the Corrected Sedimentation Coefficient

In a study conducted by Marvin and Hoffman-Berling on known values of 15 spherical viruses, M vs. r was plotted on a double log scale. The best line with slope 3/2 was drawn through the points. The equation derived from this line is

$$M = 1150 \left(\frac{s^{\circ}_{20,w} \, \bar{v}^{1/3}}{1 - \bar{v}} \right)^{3/2} \tag{5.52}$$

The value for the molecular weight of phage fr using the above equation was 3.9 × 10[6]. This value compares well with that of 4.1 × 10[6] calculated from sedimentation and viscosity data using the Scheraga-Mandelkern equation, and with the value of 4.3 × 10[6] calculated from the Svedberg equation, incorporating the sedimentation and diffusion coefficients.

An equation for M for small spherical viruses was also derived in terms of the sedimentation coefficient and the nucleic acid content. An approximate value of the partial specific volume can be obtained from the nucleic acid content if one assumes values $\bar{v}_n$ and $\bar{v}_p$ for the nucleic acid and protein:

$$\bar{v} - n\bar{v}_n + \left(1 - n\right)\bar{v}_p \tag{5.53}$$

where n is the weight fraction nucleic acid. For the known viruses noted above, nucleic acid content was plotted against $\bar{v}$. The line that best fits the points gave $\bar{v}n = 0.53$ cc/g, $\bar{v}p = 0.75$ cc/g. For $\bar{v}$ between 0.68 and 0.72, the empirical equation

$$\bar{v}^{1/3} / 1 - \bar{v} = 3.55 - 2.5n \tag{5.54}$$

fits the data to within 1%. Combining Equations 5.52 and 5.54 gives

$$M = 1150 \left[s^{\circ}_{20,w} \left(3.55 - 2.5n \right) \right]^{3/2} \tag{5.55}$$

for the molecular weight of a spherical virus in terms of the sedimentation coefficient and the nucleic acid content.

Incorporating S and D values in the Svedberg equation, Pitout et al.[400] obtained a molecular weight value of 54.3 × 10[6] for the bacteriophage PL 25. The molecular weight of phage PL 25 was correlated with known molecular weight values of 10 other phages. A linear relationship, on the log scale, was found between the corrected sedimentation coefficient and the molecular weight of the phages. An empirical equation relating $S^{0}_{20,w}$ and M was derived:

$$s^{\circ}_{20,w} = 1{,}114 \times 10^{-3} \times M^{0.729} \tag{5.56}$$

Substituting the sedimentation coefficient value of 485 and the partial specific volume value of 0.68 into the equation derived by Marvin and Hoffman-Berling, an M value of 56.0×10^6 was obtained for the PL 25 phage. However, Pitout et al. observed that one phage molecular weight value deviated from the linear relationship with $S_{20,w}^0$. The M value for the slow form of T2 phage did not fall on the line constructed graphically. From electron microscope analysis, the slow form of T2 has a longer head than the fast form. This morphological feature is probably the cause of the discrepancy noted for the slow form of T2. The observation points out the requirement that in the use of empirical equations for molecular weight calculations, it would be prudent to check the virus sample for anomalous characteristics.

The molecular weight of the Hyphomicrobium bacteriophage Hyϕ30[401] was determined from its $s_{20,w}$ value, using the equation of Pitout et al. The sedimentation coefficient of Hyϕ30 was 492 S and a molecular weight of 55.4×10^6 was calculated from the empirical equation. The bacteriophage sedimented as a single symmetrical peak, indicating that the phage preparation was homogeneous.[321]

5.5   Sedimentation Field Flow Fractionation

The sedimentation field flow fractionation (SdFFF) technique, described in Chapter 3 for the purification of viruses, can also be used to obtain the molecular weight value.[402] Viral mass is determined mainly by the measurement of the retention of virus particles in an SdFFF channel followed by the theoretical interpretation of the measured retention.

5.5.1   Retention Theory[403]

In SdFFF the flow channel is accommodated inside a centrifuge basket and is spun with an angular velocity ω. The particles within the channel experience a force F(r) given by

$$F\left(r\right)=\left(Q_s-Q\right)V\omega^2 r \tag{5.57}$$

where Q_s is the solute or particle density, Q is the solvent density, V is the particle volume, and $\omega^2 r$ is the centrifugal acceleration at radius r. Expressing V in terms of molecular weight M and particle density Q_s, and using the symbol G for the acceleration $\omega^2 r$, which can be considered constant across the narrow channel, gives

$$F=\left(MG/N\right)\left(1-Q/Q_s\right) \tag{5.58}$$

where N is Avogadro's number. The drift velocity, U, for a particle with friction coefficient f is

$$U=\frac{F}{f}=\left(MG/Nf\right)\left(1-Q/Q_s\right) \tag{5.59}$$

The friction coefficient f is given by the Einstein equation:

$$f=\frac{kT}{D} \tag{5.60}$$

where k is Boltzmann's constant, T is the absolute temperature of the system, and D is the particle's diffusion coefficient.

The key retention parameter in SdFFF is the dimensionless ratio $\lambda = \ell/w$, where w is the channel thickness and ℓ is the effective thickness of the solute layer, $\ell = D/|U|$. Thus,

$$\lambda = D\big/\big|U\big|w \qquad\qquad (5.61)$$

In SdFFF, λ plays a key role in that it links together the experimental measurement of retention volume ($\lambda = V^o/6V_r$ where V^o is the channel volume and V_r is the retention volume) and the physical chemical parameters which give rise to $|U|$ and D in Equation 5.61.

Substituting Equation 5.60 into Equation 5.59 and the latter into Equation 5.61 gives

$$\lambda = \frac{RT}{\left|GMw\left(1-Q\big/Q_s\right)\right|} \qquad\qquad (5.62)$$

where R is the gas constant, Nk. From Equation 5.62 is obtained the equation from which the true molecular weight M can be obtained, provided that the particle density, Q_s, is known:

$$M = \frac{RTQ_s}{\left|GM\lambda\left(Q_s-Q\right)\right|} = \frac{RTQ_s}{Gw\,\lambda\,|\Delta Q|} \qquad\qquad (5.63)$$

5.5.2   Effective Molecular Weight

Rearranging Equation 5.63,

$$M = \frac{|\Delta Q|}{Q_s} = M' = \frac{RT}{Gw\,\lambda} \qquad\qquad (5.64)$$

where M' is the effective molecular weight, i.e., the absolute value of the particle mass in Daltons minus the mass of displaced fluid. The value of M' can be obtained from SdFFF experiments without the benefit of known density values.

5.5.3   Molecular Weight and Density

Using Equation 5.63, both the density and mass of solute particles can be obtained by running experiments with different solvent densities. This is done by converting Equation 5.63 into the equation

$$Q = Q_s \pm \frac{Q_s}{M} \cdot M' \qquad\qquad (5.65)$$

Equation 5.65 shows that a plot of Q vs. M' gives a straight line, the intercept being equal to solute density Q_s and the slope equal to Q_s/M. The slope is positive when $Q_s < Q$ and negative when $Q_s > Q$.

SdFFF has been used to obtain the molecular weight value of a number of viruses. In addition, a molecular weight value has been resolved for the Creutzfeldt-Jakob disease agent (Table 5.13).

5.6   Obtaining the Molecular Weight by Combining Sedimentation Experiments with Other Technologies

As noted above, obtaining the translational diffusion coefficient by dynamic laser light scattering has greatly facilitated the computing of the molecular weight by the Svedberg equation. Reports are also available where combined technologies are used in the same experiment. Harding and coworkers combined gel permeation chromatography with low-speed sedimentation equilibrium to obtain the molecular

TABLE 5.13   Molecular Weight Values of Infectious Agents Obtained
From Sedimentation Field Flow Fractionation

Infectious Agent	M (millions)	Ref.
Creutzfeld-Jakob agent	15	405
fd phage	11.5	406
Gypsy moth nuclear polyhedrosis virus: single virus rod	968	407
PBCV (virus from *Chlorella*-like algae of *Paramecium bursaria*)	1046	408
φX 174	6.2	406
T2	180	406
T4D	317	404

weight distribution of macromolecules. The elution profile of a macromolecular solution from gel permeation chromatography was calibrated by extraction of a small number of narrow fractions from the eluate. Weight-average molecular weights of the fractions were determined by low-speed sedimentation equilibrium.[409,410]

A fluorescent detection system for the analytical ultracentrifuge and its application to proteins, nucleic acids, viroids, and viruses has been reported.[411]

The fluorescence detection system was developed for the Model E analytical ultracentrifuge, and was mounted in place of the conventional schlieren optics. The principle of the fluorescence system is that of a scanning microscope, in which only a very small area of the sample is illuminated by an electron beam or a laser beam, and the intensity of the emitted or reflected radiation is recorded under a wide angle of observation. An argon ion laser is used for fluorescence excitation. Sedimentation profiles were recorded with the fluorescence detection system.

Biosolute concentrations which are lower by up to four orders of magnitude as compared to conventional UV absorption are sufficient for studies with fluorescence detection.

Bovine rotavirus, cucumber mosaic virus, and TMV were studied. The combination of analytical ultracentrifugation and fluorescent detection permits the determination of a whole set of physical properties, such as buoyant density, molecular weight, s value, and associated parameters. In addition, a new method for virus detection with fluorescent antibodies in CsCl density gradient centrifugation was developed. The purpose was to characterize viruses without extended purification procedures.

6

Mass-Molecular Weight Determinations from Scattering Studies

6.1 Small Angle X-Ray Scattering (SAXS)

The overall structure of biological macromolecules in solution may be studied by scattering techniques. Scattering views structures in random orientations to a nominal structural resolution of about 2 to 4 nm in a Q range between approximately 0.05 and 3 nm. The symbol Q is a measure of the scattering angle and is equal to $4\pi\sin\theta/\lambda$. The scattering angle is equal to 2θ and λ is equal to the wavelength. Analyses of the scattering curve I(Q) measured over a range of Q leads to the molecular weight and the degree of oligomerization, the overall radius of gyration R_G, and the maximum dimension of the macromolecule.[412]

Applied to the study of globular proteins[413] and viruses[414] in the late 1940s and early 1950s, small angle X-ray scattering is capable of yielding the molecular weight. To obtain this value two auxiliary parameters are needed: the concentration and the partial specific volume of the protein.

6.1.1 Theory[415]

In 1915 Debye showed that the angular dependence of the scattering, the scattering envelope, from a particle of any shape, averaged over all orientations, is given by:[416]

$$I_{scat}\left(s\right) = \sum_{m=1}^{N} f_m \sum_{n=1}^{N} f_n \frac{\sin 2\pi s r_{nm}}{2\pi s r_{nm}}$$

$$s = \left(2/\lambda\right)\sin\theta$$

(6.1)

where N is the total number of scattering elements in the particle, f_m and f_n are the scattering factors of any pair of scattering element, r_{nm} is the distance between the elements n and m, λ is the wavelength of the radiation, and 2θ is the angle between the incident and scattered beams. From the Debye equation, the angular dependence of the scattering of variously shaped bodies can be calculated by introducing specific expressions for r_{nm}, characteristic of the geometry of the particular body. The approximate shape of a scattering particle can be determined by comparing the experimental scattering envelopes with envelopes calculated for various geometric models. It is not necessary to know the shape of a particle to obtain information about its structure.

In theory, the method of X-ray scattering differs little from that of light scattering. The differences that exist arise from differences in the wavelengths of the radiations utilized. In light scattering the wavelength is of the order of 4000 Å; in SAXS it is about 1.5 Å. Both techniques are based on concentration fluctuations of the solution under analysis. In light scattering an auxiliary parameter is the refractive index increment of the macromolecular solute which is measured directly. For SAXS such a measurement is not possible, since the refractive index is practically indistinguishable from unity. In order to express the concentration fluctuations, it is necessary to calculate a corresponding quantity, which in SAXS is the electron density — the number of electrons per unit volume. This value can be obtained from the chemical composition of the solution components.[415]

The electromagnetic theory basic to SAXS is also basic to X-ray diffraction, as the two techniques are founded on the same phenomena. X-ray diffraction results from destructive and constructive interferences in scattered radiation. This results in discrete spots or bands that correspond to characteristic repeat distances within an ordered structure, such as a crystal or virus particle. In SAXS the scattered radiation is diffuse and generally a function of angle. X-ray diffraction reflections usually correspond to small interatomic distances and are found at higher angles. SAXS corresponds principally to molecular dimensions and is concentrated mostly within a cone a few degrees from the incident beam. There is an intermediate region, 2 to 5°, in which the internal order of macromolecules begins to manifest itself. This leads to the appearance of secondary maxima and minima superimposed on the scattering curve. The result is a wavelike appearance of the angular dependence of scattering at these higher angles. The positions of such fluctuations in the scattering curve are very useful in assigning structural models to particular macromolecules.[415]

In 1939 Guinier demonstrated that scattering yields a characteristic geometric parameter of any particle which is independent of any assumption regarding its shape. This geometric parameter, the radius of gyration, R_g, is defined as the root-mean square of the distances of all the electrons of the particle from its center of electronic mass.[417]

In the case of isotropic particles, Guinier showed that for a point source radiation, and for small values of the product $R_g s$,

$$i_n(s) = i_n(0)\left(1 - \frac{4}{3}\pi^2 R_G^2 s^2 + \cdots\right) \approx i_n(0)\exp\left(-\frac{4}{3}\pi^2\left(R_G\right)^2 s^2\right) \qquad (6.2)$$

where $i_n(s)$ is the scattered intensity at angle 2θ corresponding to a given value of s, normalized to the energy of the incident beam, i.e., referred to the scattering produced by a single electron under the identical conditions, $i_n(0)$ is the normalized intensity extrapolated to zero angle. At very low angles, a plot of log $i_n(s)$ vs. s^2 gives a straight line, the slope of which is $(4/3)\pi^2 R_G^2$. The intercept, $i_n(0)$, is proportional to the square of the molecular weight (Figure 6.1).[417]

In SAXS, the utilization of a point source is rarely possible. In practice the geometry for the source is one defined by the use of narrow slits. The scattered beam is observed through slits that can be rotated around the scatterer. At very small angles of scattering, θ, the theoretical scattering intensity is $I(\theta)$ and

$$I(\phi) = Nn^2 I_e e^{-\frac{4\pi^2 R^2 \phi^2}{3\lambda^2}} \qquad (6.3)$$

where N is the total number of particles irradiated, n is the number of electrons per particle, R is the radius of gyration of the electrons in the particle, λ is the X-ray wavelength, and I_e is the normal scattering by a free electron.[418]

6.1.1.1   Instrumental Considerations[412]

Instrumental requirements are based on the irradiation of a solution of path thickness 1 to 2 mm with a collimated, monochromatized beam of X-rays, and recording the scattering pattern with a one- or two-dimensional detector linked to a minicomputer. A synchrotron X-ray camera[419,420] takes a beam of X-rays, which is emitted tangentially by the electrons circulating at relativistic speeds in the storage ring of the synchrotron. The X-ray beam is horizontally focused and monochromated to a value close to 0.15 nm in wavelength, by a Ge or Si perfect single crystal, then vertically focused by a curved mirror, and

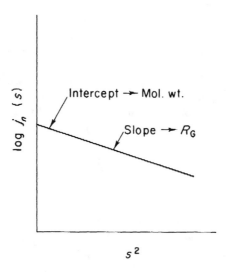

FIGURE 6.1 Information derived from a Guinier plot in small angle X-ray scattering (SAXS) . The intercept leads to the molecular weight value and the slope to the radius of gyration (R_G).

collimated by slits. Samples, 1 mm path length, with surface area 2×6 mm, and a total volume of 25 μl, are held in brass water-cooled cells with 10 to 20 μm thick mica windows or in quartz capillaries. The sample holder is aligned in the beam. Sample-detector distances, 0.5 to 5.0 m, depend on the desired measure of the scattering angle. Position-sensitive detectors are based either on a linear detector, which monitors the scattered intensities in one dimension, or on a quadrant detector, which monitors the intensities scattered in a two-dimensional angular sector (70°) of a circle with the nominal position of the main beam located at the center of the circle. Beam exposures are monitored by the use of an ion chamber positioned after the sample. The detector is interfaced with a VAX minicomputer system for data storage and online processing to assess the experimental data as it is being recorded. The camera is inside a radiation-shielded hutch, protected by safety interlocks to avoid accidental lethal exposures to users.

6.1.1.2   Sample Preparations[412]

The basic requirement is for pure, monodispersed solutions of the samples at a high enough concentration for a scattering curve to be observable in the required solute-solvent contrast. For studies on a single preparation, 0.5 ml of material at 10 mg/ml is ideal for synchrotron X-ray work. As noted, scattered intensities are proportional to the square of the molecular weight at low values of Q. Therefore, it is essential to remove all traces of aggregates prior to measurement by, for example, gel filtration and reconcentration of the samples if the sample is prone to aggregation. In the course of online data analyses during an experiment, aggregates are observed by nonlinear Guinier R_G plots, which curve upward at the lowest Q values. Such samples are not usable.

Sample dialysis prior to measurements is essential for accurate subtraction of buffer backgrounds, for which the final dialysate is employed. Slight differences in the electron density of the buffer can invalidate the subtraction. In X-ray work, the closer a protein buffer is to pure water, the higher the sample transmission becomes, and the better the results. Phosphate buffered saline ($12\ \mathrm{m}M$ phosphate, $140\ \mathrm{m}M$ NaCl, pH 7.4) is commonly used.

Sample concentrations must be accurately known prior to measurements for molecular weight. Assays for activity and radiation damage before and after scattering are required to demonstrate that the scattering data are meaningful.

In X-ray work, the instrument requires calibration. Rat tail tendons are a source of collagen fibrils, with which the Q range on the detector can be defined. Using wet, slightly stretched collagen, a diffraction spacing of 67 nm is obtained. Scattering curves are processed by subtracting the buffer curve from the sample curve and normalizing the result by dividing by the detector response curve.[412]

6.1.2   Molecular Weight[415]

Considering Equation 6.1, if each scattering element is taken as one electron in the particle, and the particle contains m electrons, then for a single particle in vacuum, it can be shown that $i_n(0) = m^2$. For J noninteracting particles per unit volume, the total scattering at $2\theta = 0°$ is $J\,m^2$. If the concentration c is expressed in mass per volume units, $J = c/m$, in vacuum,

$$\frac{i_n(0)}{c} = m \tag{6.4}$$

For a two-component system, a macromolecule in solvent, the following relation between the excess scattering of solution over solvent per unit volume $\Delta i_n(0) = i_n(0)_{solution} - i_n(0)_{solvent}$, and the fluctuations of the electron density, $\overline{\Delta\rho_s^2}$, in a volume element δV is

$$\Delta i_n(0) = \delta V \overline{\Delta\rho_s^2} \tag{6.5}$$

where $\overline{\Delta\rho^2} \equiv \overline{\rho^2} - \overline{\rho}^2$ and ρ_s is the electron density of the solution, i.e., the number of electrons per volume. Since

$$\overline{\Delta\rho_s^2} = \left(\frac{\partial\rho_s}{\partial c_e}\right)_{T,p} \overline{\Delta c_e^2} \tag{6.6}$$

where c_e is the solute concentration expressed as the ratio of the number of electrons of solute to that of the solution, i.e., a weight fraction. Application of the thermodynamic relations for concentration fluctuations results in

$$\left(\frac{\partial\rho_s}{\partial c_e}\right)_{T,p}^2 \frac{c_e(1-c_e)^2}{\Delta i_n(0)\rho_s^2} = \frac{1}{m_{app}} = \frac{1}{m}\left[1 + \frac{c_e}{RT}\left(\frac{\partial\mu^e}{\partial c_e}\right)_{T,p}(1-c_e)\right] \tag{6.7}$$

where R is the gas constant, T is the thermodynamic temperature, μ^e is the excess chemical potential of the solute, and m_{app} is an apparent mass in electrons of the particle calculated for each finite concentration of protein at which scattering measurements are made.

At constant temperature and pressure,

$$\frac{d\rho_s}{dc_e} = \rho_s\frac{(1-\rho_s\psi_2)}{(1-c_e)} \tag{6.8}$$

where ψ_2 is the electron partial specific volume of the solute. In studies of biosolutes, the measurements are performed in solution, and the intensity value used is the excess scattering of solution over solvent, $\Delta i_n(s)$.

The symbol Δ may be omitted, noting that $i_n(s)$ and $j_n(s)$ refer to the excess scattering. Combining Equations 6.4 through 6.8, expressing the derivative of the excess chemical potential, u^e, of solute with respect to concentration as the usual virial expansion, as used in light scattering, and setting $\rho_s \approx \rho_1$, since at low solute concentrations, the electron densities of solvent and solution will be the same within experimental error, leads to

$$m_{app} = i_n(0)(1 - \rho_1 \psi_2)^{-2} c_e^{-1} \tag{6.9}$$

and

$$m = m_{app} + 2Bm^2 c_e \tag{6.10}$$

where B is the second virial coefficient. Extrapolation to zero concentration of a plot of $1/m_{app}$ vs. c_e leads to m from the ordinate intercept and, with m known, to B from the slope. The molecular weight M_i is obtained from m,

$$M = mN_A/q \tag{6.11}$$

where q is the number of electrons per gram of the particle, calculated from its chemical composition, and N_A is Avogadro's number.

For a compact globular protein, the minimum R_G value can be estimated from the expression:[421]

$$\log R_G = 0.365 \log M_r - 1.342 \tag{6.12}$$

where M_r is the molecular weight. For elongated structures, R_G values can be estimated from the overall length, L, using the approximation $R_G = L/(12)^{1/2}$. L can be estimated from electron microscopy or from hydrodynamic simulations assuming a rigid rod model.

Guinier analyses of the scattering curves I(Q) at low Q gives the radius of gyration, R_G, and the forward scattered intensity, I(0)(I):[412]

$$\ln I(Q) = \ln I(0) - R_G^2 Q^2/3 \tag{6.13}$$

R_G values measure the degree of particle elongation. Molecular weights can be deduced from I(0)/c values (c = sample concentration in mg/ml) as a relative value from the X-ray data. For elongated macromolecules, the corresponding cross-sectional radius of gyration, R_{xs}, and the cross-sectional intensity at zero angle $I(Q) \times Q_{Q \to 0}$ are obtained from curve analyses in a Q range larger than that above from:[412]

$$\ln[I(Q) \cdot Q] = \{\ln[I(Q) \cdot Q]\}_{Q \to 0} - R_{xs}^2 Q^2/2 \tag{6.14}$$

After data have been reduced, it is necessary to validate the scattering curves before they can be interpreted. The curves at low Q should yield linear, reproducible Guinier plots. A concentration series of the R_G and I(0)/c data should be made.[412]

In their studies on bacteriophages f_r and R17, Zipper et al.[422] obtained small angle X-ray scattering curves from which molecular weight values were derived. In order to eliminate the influence of interparticle interferences upon the scattering curves the inner portions of the curves were measured at several phage concentrations. The reduced scattering intensities I/c were extrapolated to zero concentration in a Guinier plot.

From scattering intensity at zero angle, obtained by extrapolation of the Guinier straight line and the intensity of the primary beam, the molecular weight of the phages was determined according to the relation

$$M = \frac{K\,I_0\,a^2}{P_0 c D\left(z_1 - \bar{v}_1 Q_2\right)^2}; \quad K = \frac{1}{i_0 N_A} = 21.0 \tag{6.15}$$

where I_0 is the scattering intensity at zero angle, a = distance sample – registering plane in cm, P_0 = intensity of primary beam, c = concentration of the phage in g/ml as determined by an optical concentration measurement at 260 nm using the extinction coefficient $e = 8.03$, D = thickness of the sample in cm, z_1 = moles of electrons per gram of dissolved phage, i.e., 0.529, calculated from the components assuming a protein fraction of 70%, $\bar{v}_1$ = particle specific volume of the phage 0.673 cm³/g, Q_2 = electron density of the solvent in moles of electrons per cm³ of solvent, i_0 = Thomson's constant, i.e., the scattering of the electron, and N_A = Avogadro's number. The values of molecular weight were 3.62×10^6 and 3.7×10^6 for phages fr and R17, respectively.

In the Guinier plot, the slope of the innermost portion of the scattering curve yielded the radius of gyration, according to the relation

$$R_G = 0.644 \sqrt{\tan\alpha} \ \text{Å} \tag{6.16}$$

A value of 105.2 Å for the radius of gyration was calculated for both phages.

6.2   Small Angle Neutron Scattering (SANS)

Typically, 100 to 500 µg of solute are required for an experiment. Compared to light or X-ray radiation, neutrons have advantages that justify their use even though neutron sources are not commonly available. All these advantages stem from the particularly suitable wavelength spectrum available ($1 < \lambda < 30$ Å), the low energy of radiation, and the isotopic dependence of scattering lengths:[423]

1. Neutrons have a very wide range of application. Neutron scattering has been used to determine molecular weights in the range 10^4 to 10^9.
2. They do not inflict radiation damage on the sample. Considerable damage is caused to biological material by X-rays.
3. Neutron experiments in water solvent are not sensitive to assumptions about the partial specific volume of the particles.
4. They are not sensitive to dust that can contaminate the solutions. Dust particles cause problems in light-scattering experiments because their size is comparable to the wavelength of light and because they cause much scatter.
5. The use of D_2O-containing solvent enriches the application of neutron scattering. Because of the high signal-to-background ratio in D_2O, experiments can be performed on very low concentrations of material. Separate information on the different components of a multicomponent particle can be obtained using contrast variation by changing the water:D_2O ratio in the solvents. For example, the amount of protein and nucleic acid in a virus can be determined separately.

Neutron-scattering experiments also have certain practical advantages. Special sample containers are not required; measurements are made in standard spectrophotometer quartz cells whose optical pathlength is known, using concentrations in the 1 to 10 mg/ml range. Sample volumes are of the order 100 to 200 µl. The entire sample is observed, minimizing artifacts owing to the cell walls. Because there is no radiation damage, biochemical or functional tests may be performed on aliquots both before and after the experiment. The sample cells need not be sealed, and no special precautions are required against contamination by dust. Thus, titrations or a change in solvent conditions can be done very easily, and the measurement repeated in the same cell. The measurement itself usually takes less than 1 h.

The difficulties associated with the method are those arising from the sample and those from the neutron technique. In the first group are included the requirements of purity and monodispersity. There are no tests for monodispersity that are more sensitive than a scattering experiment. If aggregation is suspected, the scattering curve should be observed in as wide a range of concentration and buffer conditions as possible. Uncertainty of the chemical composition of the sample will add to the uncertainty of the result.

The main difficulty of the technique is that the very large incoherent scattering in water buffer, which is also used for calibration, leads to low signal-to-background ratios for low-molecular-weight particles in small concentrations, 3% to 5% for a protein of M_r 10,000 at 5 mg/ml. In D_2O solvents there may be uncertainty of H-D exchange, the sensitivity of the scattered intensity to specific volume, and the higher probability of aggregation for certain particles. In general, the method is best adapted for particles for M_r of 100,000 or more, thus making it particularly suitable for viruses.[423]

6.2.1 Experimental Apparatus: Small Angle Neutron Diffractometer[424]

A "cold" neutron beam — a beam with a higher proportion of long wavelength neutrons, usually obtained from a nuclear reactor — is directed by a series of guides to a monochromator. The guides are usually parallel-sided rectangular tubes of nickel-plated glass. The monochromator may be a crystal oriented to the beam such that only one wavelength with a narrow distribution is selected. More flexibility is obtained by using a velocity selector, consisting of a series of slotted neutron-absorbing disks. The disks are arranged on a common axis such that the slots are displaced helically from one another. By varying the rotational speed of the plates, only a narrow wavelength range of neutrons has sufficient velocity to pass unimpeded through the monochromator (Figure 6.2).

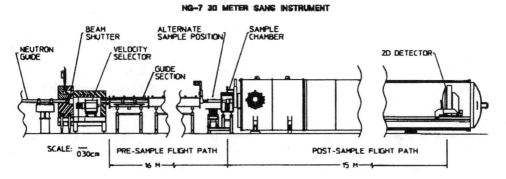

FIGURE 6.2 Neutron guide (NG) –7. A 30 m small angle neutron scattering (SANS) instrument at the Cold Neutron Research Facility of the U.S. Department of Commerce National Institute of Standards and Technology (NIST). Operating characteristics of the instrument are given in Table 6.1. (Courtesy: National Institute of Standards and Technology, Gaithersburg, MD.)

Another series of movable guides brings the neutron beam onto the sample position. After the sample position, the scattered neutrons pass down a flight tube to the detector. Both flight tube and guide tubes are evacuated to reduce air scattering. Sample positions are usually at ambient conditions and sufficiently spacious to accommodate a variety of experimental setups. The ends of the guide and flight tube are sealed with quartz windows which transmit neutrons with little scattering (Table 6.1).

The detector is generally a position-sensitive, two-dimensional detector and may be placed at distances selected by the user from the sample position, thus defining the accessible Q range. As noted in the previous section, $Q = 4\pi \sin \theta/\lambda$; 2θ = scattering angle; λ = wavelength. Q is a measure of the scattering angle and is equal to $2\pi/d$, where d is the diffraction spacing specified in Bragg's law of diffraction: $\gamma = 2 d \sin \theta$.[412]

TABLE 6.1   30 m Small Angle Neutron Scattering Characteristics

Source size	5×5 cm
Wavelength range	0.4–2.0 nm (velocity selector)
Wavelength resolution	7%–30% (continuously tunable)
Q-Range	0.01–6 nm^{-1} (CHRNS-SANS)
	0.01–10 nm^{-1} (NIST/EXXON/Univ. of Minn. SANS)
Sample size	0.5–2.5 cm
Expected flux at sample	10^3 to 10^6 n/cm^2·s depending on slit size, wavelength, and source to sample distance

Courtesy: National Institute of Standards and Technology, Gaithersburg, MD.

The detector is connected to a computer, which records the received counts and any counts on monitor detectors placed in the beam. On completion of a measurement, the operating system automatically records the data.[424]

6.2.2   Molecular Weight[423]

The determination of molecular weights by neutron scattering has been analyzed by Jacrot and Zaccai[423] and is presented here.

In a solution of particles, neutrons are scattered by all atomic nuclei — in the solvent as well as in the solute. For N identical particles in solution,

$$I(0) = N \left(\Sigma b - \rho_s V \right)^2 A \qquad (6.17)$$

where I(0) is the coherent intensity scattered at zero angle, Σb is the sum of coherent scattering lengths over all nuclei in each particle, V is the volume of the particle, ρ_s is the scattering density (scattering length per unit volume) of the solvent, and A is a factor that includes the incident beam intensity, the size of the sample, its transmission, and all geometrical parameters of the experiment.

Equation 6.17 is strictly valid for a sample that is sufficiently dilute and thin so that scattered waves from different particles do not interfere with each other. When a relatively high concentration is required in order to have a reasonable signal, one should measure different samples and extrapolate the results to zero concentration.

Equation 6.17 is valid for any radiation, provided the appropriate scattering lengths (obtained from tables) are used. It allows a determination of the molecular weight of a particle because, for a given composition, Σb and V are proportional to M_r, the molecular weight, and for a given concentration C (in mg/ml) N is inversely proportional to M_r. Then,

$$I(0)/C \simeq \text{constant} \times \frac{1}{M_r} \times M_r^2 = \text{constant} \times M_r \qquad (6.18)$$

Incoherent scattering gives rise to a background which is angle independent at small angle, and which must be subtracted from the signal measured. In practice, because solutions are sufficiently dilute (less than 1%), the incoherent scattering is considered to arise entirely from the solvent, so that a measurement with a sample containing only solvent allows its determination. The incoherent scattering is of the same order of intensity as I(0) for a water solvent, and much smaller for a D$_2$O solvent. After subtraction of the solvent background, I(0) is determined by extrapolating the coherent intensity to zero angle, using the Guinier approximation.[425] The relation between I(0) and molecular weight, as presented,[423] does not require calibration against a standard.

Following Equation 6.17, the determination of molecular weight requires the following information:

1. The factors A and N
2. The relation between Σb and molecular weight, M_r
3. The relation between V and M_r, namely the partial specific volume, $\bar{v}$
4. The concentration C of the sample

6.2.2.1 The Factors A and N

The factor A is given by

$$A = T_s I_0 \Omega \tag{6.19}$$

where T_s is the transmission of the sample, I_0 is the number of neutrons per cm^2 per second in the incident beam; Ω is a geometrical term, which takes into account the angular aperture of the incident beam, and the solid angle used for detection.

The incoherent scattering of water is relatively strong and has negligible angular dependence. The exact number of scattered neutrons is wavelength-dependent and can be measured by observing the transmission T of the direct beam by the water sample. When a beam of I_0 neutrons cm^{-2} s^{-1} is incident on a water sample of cross-sectional area S cm^2, it will transmit $TI_0 S$ neutrons per second. The remainder, $(1 - T)I_0 S$ neutrons per second, are scattered incoherently. Assuming the scattering to be isotropic, the number of neutrons scattered in a solid angle is given by

$$I_{inc} = \frac{1}{4\pi}\left(1-T\right)I_0 S\Omega \left(\text{neutrons}/\text{s}\right) \tag{6.20}$$

The assumption that the incoherent scattering is isotropic is a good approximation only for neutrons of wavelength greater than or equal to 10 Å. Because of the recoil of atomic nuclei in collision with neutrons, incoherent scattering is strongest in the forward direction and falls off with angle. This effect depends on wavelength, λ, and the anisotropy in the scattering distribution is most pronounced at short wavelengths. It is expressed by the parameter f, defined as

$$f = \left(1-T\right)I_0 S\Omega / I_{inc}\left(0\right)4\pi \tag{6.21}$$

where $I_{inc}(0)$ is the incoherent scattering of water in the forward direction.

The value of I_{inc} is measured by using a water sample after subtraction of the contribution of the container. When the measurement is done under identical experimental conditions as for the sample containing the particles, the factor $I_0 \Omega$ can be evaluated so that A can be written

$$A = fT_s I_{inc}\left(0\right)4\pi / \left(1-T\right) \tag{6.22}$$

N is related to the concentration C and M_r by

$$N = CN_A St / M_r \tag{6.23}$$

where t is the pathlength in the sample, S is its cross-sectional area, and N_A is Avogadro's number. If C is in g l^{-1} and St is in cm^3, then

$$N = CN_A St / M_r \times 10^{-3} \tag{6.24}$$

Combining these relations with Equation 6.17 the following expression for I(0) is obtained:

$$\frac{I(0)}{I_{inc}(0)} = f\frac{4\pi T_s}{(1-T)}M_r N_A t \times 10^{-3}\left(\frac{1}{M_r}\left(\Sigma b - \rho_s V\right)\right)^2 \tag{6.25}$$

The units are as follows: I(0), I_{inc} are in neutrons s^{-1} in the same solid angle; C is in g l^{-1}; Ts,T are the measured transmissions of the sample and water, respectively; M_r is a molecular weight ratio, and is therefore dimensionless; t is in cm; and $1/M_r$ $(\Sigma b - \rho_s V)$ is in cm per unit M_r.

All other terms being constants or readily measured, I(0)/C is proportional to M_p when the relative scattering length $\Sigma b/M$ and the specific volume of a particle are known.

6.2.2.2 Relation Between Σb and Molecular Weight

For proteins, the value of $\Sigma b/M_r$ can be calculated precisely when its amino acid composition is known. The scattering lengths of amino acid residues and the scattering length per unit M_r for several proteins can be obtained from tables. To a very good approximation, Σb is proportional to M_p and a very good estimate of M_r can be made for a protein whose amino acid composition is undetermined. In the case of a nucleic acid, the molecular weight and scattering amplitude are practically proportional to each other. The scattering lengths of nucleotides are also given in tables. For a two-component system, such as a nucleoprotein, the resulting coherent scattering length is the sum of that of the two components. When the relative proportions of the components are known, this value is readily calculated.

6.2.2.3 Relation Between V and M_r

The particle volume V, implied in Equation 6.17, is that which excludes the solvent. The molecular weight and V are related by

$$V = M_r \bar{v}/N_A \tag{6.26}$$

where N_A is Avogadro's number and $\bar{v}$ is the partial specific volume of the particle. In neutron scattering, an error of 5% in $\bar{v}$ leads to an error of only 1.5% in the determination of M_r. This is a notable advantage of neutrons over X-rays, where an error in $\bar{v}$ leads to a much larger error in the molecular weight determination. A similar argument applies to nucleic acids. The determination of M_r in water solution, therefore, is not sensitive to the assumption for $\bar{v}$, and mean values can be taken for a protein and for a nucleic acid.

6.2.2.4 Multicomponent Particles

To deal with a particle made of several components of different chemical nature (e.g., protein and nucleic acid, protein, nucleic acid, and lipids), Equation 6.25 is generalized. If M_i is the molecular weight of component i, and M_r the total molecular weight of the particle, then,

$$\frac{I(0)}{I_{inc}(0)C} = f\frac{4\pi T_s}{(1-T)}M_r N_A \times 10^{-3}t\left\{\sum_i\left(\frac{(\Sigma b)_i - \rho_s V_i}{M_i}\right)\frac{M_i}{\Sigma M_i}\right\}^2 \tag{6.27}$$

with $M_r = \Sigma M_i$.

By manipulating the contrast by changing either ρ, the scattering length density of the macromolecule, or ρ_s, the scattering length density of the solvent, the intensity of the scattered neutrons at Q = 0 can be increased or decreased, depending upon the value of $\Delta\rho$. If ρ_s is changed by adjusting the ratio of H_2O to D_2O in the solvent, then $\Delta\rho$ varies linearly with the concentration of D_2O (% D_2O) in the solvent. A plot of $\sqrt{I(0)}$ vs. % D_2O yields a straight line which crosses the x-axis at the point where the neutron intensity vanishes. This is

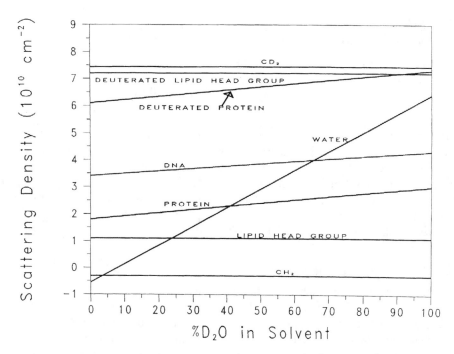

FIGURE 6.3 Scattering density of different chemical components of biological complexes as a function of the D_2O/H_2O mixture in which they are immersed. The scattering densities are calculated assuming that all labile hydrogens (e.g., those attached to O or N) do in fact exchange with the solvent, except in proteins, where a figure of 80% exchange is assumed. Horizontal lines (e.g., CH_2) are for compounds having no labile hydrogen. (Courtesy: National Institute of Standards and Technology, Gaithersburg, MD.)

the match point of the molecule. This match point is especially useful for composite systems, such as protein-DNA or lipid-protein complexes. Using the method of "contrast variation," the scattering from one component can be minimized by adjusting the H_2O/D_2O ratio in the solvent, thus allowing the scattering from the other component to dominate the total scattering.[426] Figure 6.3 shows the scattering length density from several components of biological complexes as a function of % D_2O in the solvent.[427] The scattering density from water varies from -0.562×10^{10} cm^{-2} for H_2O to 6.4×10^{10} cm^{-2} for D_2O.[426]

In the small angle limit, the Guinier approximation for the scattered intensity,[427]

$$I(Q) \sim I(O)\exp\left(-Q^2 R_g^2 / 3\right) \tag{6.28}$$

where R_g is the radius of gyration of the macromolecule, applies in the range $QR \le 1$. Both $I(O)$ and R_g may be determined from a plot of $\ln[I(Q)]$ vs. Q^2. $I(O)$ is related to the molecular weight of the molecule and R_g to the shape. Beyond the Q region, where Equation 6.28 is valid, $I(Q)$ must be compared to model curves in order to gain further information about the macromolecular structure. SANS has been useful for rapid measurement of the radius of gyration and molecular weight of macromolecules in solution. Rapid characterization can be provided in solution in a non-"invasive" manner.[426]

Equation 6.27 can be applied when, for each chemical component, the scattering length per unit mass is known in water, as well as in solvents containing D_2O. When only the concentration of the particles is known, Equation 6.27 can be used to determine the molecular weight of each component, by experiments in water and in solvents of scattering density identical to that of each component. It is often the case that the weight ratio $M_i/\Sigma M_i$ of the two components is known from chemical analysis. Equation 6.27 will give the total molecular weight of the particle directly. Contrast variation can be used, as above, to verify the value of the weight ratio.[426]

6.2.2.5 Rodlike Particles[423]

For rodlike particles one can define a mass per unit length which can also be measured. The relation used is

$$\lim_{Q \to 0} QI(Q) = \pi \left(1 - \rho_s \frac{V}{B} \right)^2 B\mu A \qquad (6.29)$$

Q is the scattering vector $4\pi \sin\theta/\lambda$, where 2θ is the scattering angle and λ the wavelength; A is as described; $\mu = B/L$ is the scattering length per Å, and ρ_s is as described. The units of Q are Å^{-1}.

Equation 6.29 is equivalent to Equation 6.17 and can be used in an equivalent way to derive the molecular weight per unit length. Such a determination does not require a monodisperse sample. The rods can have different lengths provided they all have the same mass per unit length. If the sample is monodisperse, one can measure both the molecular weight and the mass per unit length of the rods. This requires very different ranges of scattering vector and can be done much more easily with neutrons than with X-rays.

6.2.2.6 Concentration of the Sample[423]

Determination of the concentration is the limiting factor in accuracy in all scattering methods for determining molecular weights. Because the absolute concentration is required, the absorbance in the UV wavelength must be calibrated by methods that give absolute value: carbon and nitrogen content or amino acid analysis with a reference, for example. For proteins whose extinction coefficients at 280 nm are known, there is another advantage in using neutrons. The experiments are done in 1-mm quartz cells with concentrations close to 5 g/l. This corresponds to a UV transmission of about 0.5, so that the concentration measurement can be obtained directly in the same cell that is used for the scattering experiment.

Small angle neutron scattering from solutions of small RNA viruses has been used to study protein-nucleic acid organization. The viruses analyzed were turnip yellow mosaic virus, Southern bean mottle virus, cucumber mosaic virus, bromegrass mosaic virus, and MS2 phage. The results strongly suggested the existence of large central holes in all these viruses, with the possible exception of turnip yellow mosaic virus, where the hole, if any, is much smaller.[428]

The first application of the contrast variation technique to small spherical plant viruses showed the RNA and protein not seen by X-ray crystallography. In tomato bushy stunt virus, there is a bilobal radial distribution of protein[429] and a more homogeneous distribution in Southern bean mosaic virus.[430]

The method has also been applied to the much larger animal viruses, such as adenovirus[431] and influenza virus.[432] In these cases the particles were originally less well characterized than the smaller plant and phage viruses. Basic parameters such as molecular weight were poorly known. Neutron scattering allowed the molecular weight of adenovirus to be determined as well as the radial distribution of DNA and protein. In the case of influenza virus, a lipid-containing RNA virus, the molecular weight was redetermined and at the same time the chemical composition, in particular the protein/lipid ratio. The position of the lipid bilayer was also determined precisely.[432]

The molecular weight, *in situ*, of the RNA and of the protein in satellite necrosis virus has been determined by neutron scattering.[433] The molecular weight of Semliki Forest virus[434] determined by this procedure gave a value of 40 million, and for frog virus 3, a molecular weight of 520 million was determined.[435]

6.3 Classical Light Scattering

The science of light scattering was first described by Lord Raleigh at the end of the 19th century. Classical light scattering provides an absolute method for the determination of molecular weight of macromolecules.

As opposed to "dynamic" light scattering (see Chapter 5), classical light scattering means the measurement of the total or time-integrated intensity scattered by a macromolecular solution compared with the incident intensity for a range of concentrations and/or angles. This information can be used to obtain the molecular weight M, gross conformation from the radius of gyration, and thermodynamic nonideality parameters, particularly the thermodynamic second virial coefficient B (also known as A), which can also yield useful information on molecular conformation. Classical light scattering has a variety of synonyms: total intensity light scattering, differential light scattering, traditional light scattering, static light scattering, integrated light scattering, or simply light scattering.[437]

Trace amounts of large particle contaminants in light scattering experiments give serious errors in molecular weights and radii of gyration. This is known as the "dust problem."[438] However, two developments have made the light scattering technique a valuable tool for biochemists and virologists: (1) the use of laser light sources, providing high collimation, intensity, and monochromaticity, (2) the coupling of gel permeation chromatography (GPC) systems online to a light scattering photometer facilitates the analysis of polydisperse materials and provides a very effective online clarification system from dust and other large contaminants.[437]

6.3.1 Theory[438]

Light passing through any medium remains undeflected as long as the polarizability and density are uniform. Whenever variations in these factors takes place, light is scattered in all directions. The shape of the scattering envelope depends on the size and shape of the particles and on the wavelength of the electromagnetic radiation employed.

If particles are immersed in a medium of polarizability, α_o, the observed increase in scattering when the particles are introduced into the medium is the result of the excess polarizability of the particles over that of the medium. All the molecules in the solution are in constant thermal motion, so that there will be constant concentration and density fluctuations in a volume element, δV, if it is small enough. Upon examining the volume element over a period of time, its total polarizability will be found to fluctuate. The scattering in solution is proportional to the time average of the fluctuations in the polarizability within a volume element. The total scattering from all volume elements from concentration fluctuations of the macromolecules is

$$R_\theta = \frac{8\pi^4 V \delta V \overline{\Delta C^2}}{\lambda_0} \left(\frac{\partial \alpha}{\partial C} \right)_{T,p}^2 \qquad (6.30)$$

where R_θ is the Rayleigh ratio and is related to the turbidity by $r = 16\pi R_\theta/3$, C is the concentration of the macromolecule in grams per milliliter, α is the polarizability, and λ_0 is the wavelength of the light *in vacuo*.

The polarizability may be related to the refractive index, n, by the Maxwell and Lorenz equations

$$\alpha = \left(n^2 - 1 \right) \big/ 4\pi \qquad (6.31)$$

and Equations 6.30 and 6.31 are combined with the thermodynamic relationship

$$\delta V \overline{\Delta C^2} = \frac{-kTC_2 \overline{v}_1}{\left(\dfrac{\partial \mu_1}{\partial C_2} \right)_{T,p}} \qquad (6.32)$$

where k is Boltzmann's constant, $\bar{v}_1$ is the partial molal volume of the solvent, and μ_i the chemical potential. Thus, in dilute solution, the light scattering equation of Debye is obtained and may be presented in three commonly used forms for a two-component system:

$$\frac{KC_2}{R_\theta} = \frac{1}{M_2}\left[1+\left(\frac{\partial \ln\gamma_2}{\partial C_2}\right)C_2\right] \tag{6.33}$$

$$\frac{KC_2}{R_\theta} = \frac{1}{M_2}\left[1+\frac{C_2}{RT}\frac{\partial\mu_2^{(e)}}{\partial C_2}\right] \tag{6.34}$$

$$\frac{HC_2}{\Delta\tau} = \frac{KC_2}{R_\theta} = \frac{1}{M_2}\left[1+2BC_2+3CC_2^2+\cdots\right] \tag{6.35}$$

where $K = 2\pi^2 n^2 (\partial n/\partial C_2)_{T,P}^2/N_A\lambda^4$, $H = 16\pi K/3$, γ_2 is the activity coefficient of the macromolecules, $\mu_2^{(e)} = RT\ln\tau^2$ is its excess chemical potential, B, C, etc. are the second, third, and higher virial coefficients, $\Delta\tau$ is the excess turbidity of solution over pure solvent, and N_A is Avogadro's number.

When KC_2/R_θ or $HC_2/\Delta\gamma$ is plotted as a function of C_2, a curve is obtained, the intercept of which is the reciprocal of the weight-average molecular weight. The slope is 2B, the second virial coefficient. The experimental quantities required are the refractive index increment of the macromolecule $(\partial n/\partial C_2)$, and the turbidities of the solvent and of solutions of different concentrations of the macromolecule.

6.3.2   Radius of Gyration[438]

In light scattering when the dimensions of a particle become comparable to the wavelength of the incident radiation, interference occurs between the radiation scattered from elements within the particle. This effect becomes significant when the maximal dimension of a particle is of the order $\lambda/10$. The wavelength of the incident radiation is of the order of 4000 Å, and thus the particle dimensions must attain 400 Å in order to be resolved by light scattering.

The relationship between the angular dependence of the scattering intensity, I(h), and the radius of gyration R_G is

$$I(h) = K'M_2^2\left(1-\frac{h^2}{3}R_G^2+\cdots\right) \tag{6.36}$$

$$h = \frac{4\pi\sin(\theta/2)}{\lambda}$$

where M_2 is the molecular weight of the macromolecule, R_G is its radius of gyration, θ is the angle formed between the directions of the incident and scattered rays, and $\lambda = \lambda_0/n$, the wavelength measured in a medium of refractive index, n. This equation for light scattering is identical to Equation 6.2 (SAXS), in which the angular function is expressed by the symbol s.

In a volume V of J noninteracting particles, the total scattering is the scattering from a single particle multiplied by $J = C_2N_A/M_2$, where C_2 is the concentration of the macromolecules in grams per milliliter, and N_A is Avogardo's number. Combining this relationship with Equation 6.36 and introducing the proper optical and instrumental constants, results in

$$\frac{KC_2}{R_\theta} = \frac{1}{M_2}\left(\frac{1+h^2R_G^2}{3}-\cdots\right) = \frac{1}{M_2P(\theta)} \tag{6.37}$$

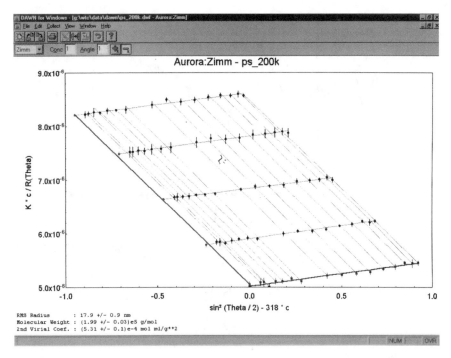

FIGURE 6.4 Zimm plot (refer to text). (Courtesy: Wyatt Technology, Santa Barbara, CA.)

If the particle is sufficiently large, its scattering intensity will vary with the angle of observation. The scattering intensity is also a function of concentration expressed through the second virial coefficient 2B.

In order to determine the second virial coefficient B, the scattering at each concentration is extrapolated to zero angle, and the determination of the radius of gyration R_G requires the extrapolation of the angular variation of scattering to zero concentration at each angle. The molecular weight calculation requires both extrapolations. These two requirements are expressed in a single equation:

$$\frac{KC_2}{R_\theta} = \frac{1}{M_2}\left(1 + 2BG_2 + \frac{16\pi^2 \sin^2\left(\theta/2\right)}{3}\frac{R_G^2}{\lambda^2}\right) \tag{6.38}$$

which is the basis for Zimm's consideration of light scattering data, the Zimm plot.[439] In such treatment, KC_2/R_θ is plotted against $\sin^2(\theta/2) + kC_2$, where k is an arbitrary constant that allows the data to be spaced conveniently, and facilitates extrapolations to zero angle and zero concentrations (Figure 6.4).

Johnson and Gabriel[440] explain Zimm plots as follows. The data points fall on a grid, producing two families of curves, one corresponding to constant concentration and the other to constant angle. The data points on the grid corresponding to a given angle are then extrapolated to zero concentration, and similarly the points at a given concentration are extrapolated to zero angle. The double extrapolation then produces two new sets of data points, one at zero angle and the other at zero concentration from which the information about the particle shape and weight is obtained. The inverse of the weight-average molecular weight is obtained from the intercept of the $\theta = 0$ curve, and the second virial coefficient B is obtained from the slope (2B = slope). The intercept of the C = 0 curve again gives the inverse of the molecular weight, and the initial slope is proportional to the radius of gyration.

Yang[441] described an alternate way of plotting Equation 6.38. $KC_2/[R_\theta \sin^2(\theta/2)]$ is plotted as a function of $1/\sin^2(\theta/2)$ with concentration, C_2, as a parameter. This plot gives a series of straight lines with a common intercept. Extrapolation of the data at each angle to zero concentration gives the weight-average molecular weight from the reciprocal of the slope. The radius of gyration is obtained from the common intercept on the ordinate. A second plot is required for the determination of the second virial coefficient, 2B. In this plot, KC_2/R_θ is plotted as a function of $1/C_2$, and the data at each concentration are extrapolated

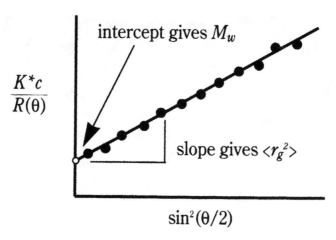

FIGURE 6.5   Plot used in classical light scattering to obtain the molecular weight value (M_w) and the radius of gyration (r_g). (Courtesy: Wyatt Technology, Santa Barbara, CA.)

to zero angle. The common intercept gives the value of 2B; and the slope at zero angle, the reciprocal of the molecular weight.[438]

6.3.3   Types of Measurement[437]

6.3.3.1   Turbidimetry

This procedure involves the measurement of the total loss of intensity by a solution through scattering, summed over the entire angular intensity envelope, compared with the intensity of the incident radiation. A good-quality spectrophotometer can perform measurements at wavelengths away from the absorption maxima. Turbidimetry is generally suitable for the measurement of molecular weights of viruses and estimating the number concentration of bacteria and bacterial spores.[442]

6.3.3.2   Low-Angle Light Scattering

In this procedure scattering measurements are performed at only one fixed small angle ($\leq 8°$). The angle is assumed low enough such that no angular correction of the scattering data is required, although the extrapolation to zero concentration of (K_C/R_θ) may be necessary. The method provides values for M and B of a system, but not R_G, since no record is made of the angular dependence of K_C/R_θ. The dust problem is very important in this procedure. However, the online coupling of a low-angle laser light scattering (LALLS) photometer to GPC has largely circumvented this problem and facilitates the measurement of molecular weight distributions for heterogeneous materials.[443,444]

6.3.3.3   Multiangle Light Scattering

This procedure can utilize a goniometer arrangement or fixed detectors at multiple angles. Obtaining measurements at multiple angles permits extrapolation of the ratio K_C/R_θ to zero $\sin^2(\theta/2)$ (Figure 6.5), which, together with an extrapolation to zero concentration, is the basis of Zimm plots. This method can yield M, B, and R_G. Plots of (K_C/R_θ) are only linear over a wide range of angles for randomly coiled macromolecules. Plots of ln (K_C/R_θ) vs. $\sin^2(\theta/2)$, Guinier plots, facilitate the angular extrapolation for globular macromolecules.[445]

Multiangle laser light scattering (MALLS) photometers have been coupled online to GPC, facilitating the analysis of heterogeneous solutions and also largely circumventing the dust problem.[446,447] Distributions of R_G, provided the maximum dimension of the macromolecule is $>\lambda/20$, as well as molecular weight can be obtained. Nonideality effects are not usually as severe for GPC-MALLS, since concentrations of volume "slices" passing through the flow cell are much smaller than the initial loading concentration. In many cases an extrapolation to zero concentration or knowledge of the second virial coefficient is not necessary to obtain a satisfactory estimate of M.

6.3.4   Experimental Considerations[437]

Solutions should be dialyzed against an appropriate buffer of defined pH and ionic strength. For scrupulously clean solutions, a 5-mW laser for a loading concentration of 3 mg/ml is sufficient for particle molecular weight as low as $\approx 40,000$.

A relatively recent commercial model, a GPC-MALLS system (Wyatt Technology, Santa Barbara, California) is available.[447] The light scattering photometer has to be "calibrated," usually with a strong Rayleigh scatterer (maximum dispersion $<\lambda/20$) whose scattering properties are known. Calibration is necessary, because the ratio of intensities of the scattered and incident beams is usually very small ($\approx 10^{-6}$). For simultaneous multiangle detection, the detectors have to be "normalized" to allow for the different scattering volumes as a function of angle and the differing responses of the detectors. A suitable concentration detector is incorporated downstream from the light scattering photometer. The volume delay between the light scattering photometer and the concentration detector needs to be accurately known. The refractive index increment at the scattering wavelength used, if not known, needs to be measured.[448] Further, if the second virial coefficient, B, is not known, and if column loading concentrations are high, the K_c/R_θ ratio needs to be evaluated as a function of concentration as well as angle, and a double extrapolation to zero angle/zero concentration can be performed using a Zimm plot.[437]

The molecular weights of bacteriophage MS2 and its RNA were determined by light scattering.[449] Measurements were made on a Brice-Phoenix scattering photometer, model 1000D (Phoenix Precision Instruments, Inc., Gardner, New York) using the 4358 Å Hg line. Solutions were first cleaned of dust by filtration through type HA Millipore filters (Millipore Filter Co., Bedford, Massachusetts).

The refractive increment of the virus was 0.191 ml/g, at 436 nm on a Brice-Phoenix differential refractometer. Concentrations were determined spectrophotometrically. The light scattering was performed at several concentrations, and within experimental error, the intercept was independent of concentration. From the intercept, the particle weight was 3.6×10^6. The viral RNA has a molecular weight of 1.05×10^6, indicating that there is one molecule of RNA per virus particle.

Overby et al.[450] also used light scattering measurements to obtain the molecular weight of MS2 and Qβ phages. The scattering envelopes of both phages were determined at 546 nm at five different concentrations. The refractive increments, determined in a differential refractometer, were 0.199 ml/g for MS2 and 0.198 ml/g for Qβ. Within experimental error, the intercepts were independent of concentration. From the intercepts, the particle molecular weights of MS2 and Qβ were 3.6×10^6 and 4.2×10^6, respectively. These values compared favorably with molecular weight values for the phages calculated from sedimentation and diffusion coefficients.

Schito et al.[325] employed Zimm plots in their light scattering studies on the DNA bacteriophage, N4. From Zimm plots, a mean particle weight of 83×10^6 was obtained. The specific refractive increment of the virus was 0.170 ml/g. Assuming the phage to be a sphere, the dissymmetry of the scattering envelope was 1.410, indicating a diameter of 660 Å.

Laser light scattering has been combined with a variety of detectors in order to enhance the precision and accuracy of the data, including the molecular weight value. Size exclusion chromatography (SEC) is used with online scattering detectors.[451,452] The molecular weight from this measurement is independent of the elution position. This characteristic makes the combination of light scattering with SEC an easy, accurate, and reliable technique.

Detectors of viscosity parameters have also been combined with laser light scattering instrumentation.[453-456] The procedure of field flow fractionation discussed in Chapter 5 has been coupled to laser light scattering (Figure 6.6). This feature gives the user separation capabilities and molecular weight determinations together.[457]

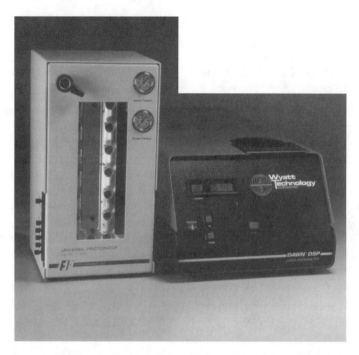

FIGURE 6.6   Coupling of a field flow fractionator to the measurement capabilities of a light scattering instrument. (Courtesy: Wyatt Technology, Santa Barbara, CA.)

In yet other studies, investigators have employed three detectors, including laser light scattering, with SEC. Wen et al.[458] used SEC with online light scattering, absorbance, and refractive index detectors for the study of proteins and their interactions. Haney et al.,[459] also used three detectors: light scattering, a viscometer, and a refractometer to obtain data on molecular weight, size, and density, giving added power to SEC in revealing structural details. All of these innovative methodologies are applicable to virus study, including molecular weight determinations.

A survey of recent instrumentation is available.[460] Most of the new instruments permit dynamic as well as classical light scattering. Table 6.2 contains values of molecular weights of viruses obtained from scattering studies.

6.4   Electron Microscopy

The procedure of quantitative electron microscopy has been utilized to obtain the mass of particle in the range 10^{-10} to 10^{-18} g.[461,462] The dry mass equivalent of every virus particle whose image is recorded on an electron micrograph can be analyzed directly by quantitative electron microscopy and the researcher provided with a range of mass values contributed by the individual particles. From analysis of the population of virus particles studied an average mass value is obtained and may be converted to the more common molecular weight value.

The simplicity of the method, allowing one to isolate viruses from small amounts of material without further purification, as was reported for herpes simplex virus,[463] is of great value. In addition, size and shape can be determined from the same micrograph and correlated with the respective particle mass.

Since its introduction as a research procedure, mass determination with the electron microscope has been extended through the use of high voltage transmission electron microscopy and quantitative scanning transmission electron microscopy. These procedures are discussed below.

TABLE 6.2   Mass-Molecular Weight (M) Values from Scattering Studies

Specific Virus	Mass (g)	M (g/mol)	Procedure
Blue green algal:			
lPP-1M	8.9×10^{-17}	5.3×10^{7}	QTEM[490]
lPP-2 isolate WA	8.5×10^{-17}	5.1×10^{7}	QTEM[490]
Cabbage moth NPV			
Inclusion body	2.5×10^{-12}	1.5×10^{12}	HVQTEM[491]
Cow pox	6.5×10^{-15}	3.9×10^{9}	QTEM[470]
Equine herpes nucleocapsid			
Intermediate density		2.3×10^{8}	QSTEM[492]
Light density		2.0×10^{8}	QSTEM[492]
European pine sawfly NPV			
Inclusion body	4.0×10^{-14}	2.4×10^{10}	HVQTEM[493]
Single rod	4.7×10^{-16}	2.8×10^{8}	QTEM[494]
Round form	4.6×10^{-16}	2.8×10^{8}	QTEM[494]
fd phage		1.6×10^{7}	QSTEM[486]
	184 Da/nm		SANS[436]
fr phage		3.6×10^{6}	SAXS[422]
Fowlpox			
Australia	6.0×10^{-15}	3.6×10^{9}	QTEM[470]
Mississippi	7.2×10^{-15}	4.4×10^{9}	QTEM[470]
Frog virus 3		5.2×10^{8}	SANS[435]
Gypsy moth NPV			
Inclusion body	3.4×10^{-12}	2.1×10^{12}	HVQTEM[495]
Single rod	1.6×10^{-15}	9.4×10^{8}	QTEM[494]
Herpes simplex			QTEM[463]
Core	2.1×10^{-16}	1.3×10^{8}	
Empty capsid	5.2×10^{-16}	3.2×10^{8}	
Full capsid	7.6×10^{-16}	4.6×10^{8}	
Enveloped nucleocapsid	1.3×10^{-15}	8.0×10^{8}	
Influenza		1.8×10^{8}	SANS[432]
MS2 phage		3.6×10^{6}	LS[449,450]
Myxoma	6.2×10^{-15}	3.8×10^{9}	QTEM[470]
N4 phage		8.3×10^{7}	LS[325]
Nudaurelia cytherea capensis		1.6×10^{7}	EM counting[482]
pf1 phage	157 Da/nm		SANS[436]
QB phage		4.2×10^{6}	LS[450]
R17 phage		3.7×10^{6}	SAXS[422]
Rabbitpox	5.3×10^{-15}	3.2×10^{9}	QTEM[470]
Semliki forest		3.5×10^{7}	QSTEM[434]
		4.0×10^{7}	SANS[434]
SV40	5.4×10^{-17}	3.3×10^{7}	QTEM[494]
Spruce budworm NPV Inclusion body	2.0×10^{-12}	1.2×10^{-12}	QTEM[493]
T2 phage	3.5×10^{-16}	2.1×10^{8}	QTEM[496]
T5 phage	1.9×10^{-16}	1.1×10^{8}	QTEM[496]
T7 phage	7.4×10^{-17}	4.5×10^{7}	QTEM[490]
Tipula iridescent	1.4×10^{-15}	8.2×10^{8}	QTEM[494]
Tobacco mosaic		4.9×10^{7}	EM counting[497]
		4.2×10^{7}	EM counting[482]
		4.1×10^{7}	LS[498,499]
		5.0×10^{7}	LS[500]
		5.1×10^{7}	LS[501]
	7.6×10^{-17}	4.6×10^{7}	QTEM[494]
		3.9×10^{7}	QSTEM[502]

TABLE 6.2 (continued)   Mass-Molecular Weight (M) Values from Scattering Studies

Specific Virus	Mass (g)	M (g/mol)	Procedure
Tomato bushy stunt		7.8×10^7	QSTEM[434]
		9.4×10^7	EM counting[481]
Vaccinia strains			
NIH	6.5×10^{-15}	3.9×10^9	QTEM[467,470]
Hamburg	5.3×10^{-15}	3.2×10^9	QTEM[467,470]
Lea (7N)	6.2×10^{-15}	3.7×10^9	QTEM[467,470]
WR	6.0×10^{-15}	3.6×10^9	QTEM[467,470]
Vesicular stomatitis		2.7×10^8	QSTEM[503]

Abbreviations: QTEM, quantitative transmission electron microscopy; NPV, nucleopolyhedrosis virus; HVQTEM, high voltage quantitative transmission electron microscopy; QSTEM,
quantitative scanning transmission electron microscopy; QTEM, quantitative transmission electron microscopy; SANS, small angle neutron scattering; SAXS, small angle X-ray scattering; LS,
light scattering.

6.4.1   Quantitative Transmission Electron Microscopy (QTEM)

The essential features of QTEM are (1) the relationship between mass per area of a particle and the
transmission recorded for its image on an electron micrograph negative, plate or film; and (2) through
integrating photometry, the conversion of photometric measurement to mass.

In the transmission electron microscope, a thin object, e.g., a virus particle, transradiated by electrons
alters initial energy and direction:[464]

$$I = I_0 \, e^{-\alpha Sw} \qquad (6.39)$$

where I_0 is the unscattered beam intensity; I is the intensity of the beam after it has passed through the
object; α is an instrumental factor determined by the operating parameters and design of the microscope;
S is the scattering cross section (cm^2) per gram of substance and is analogous to the extinction coefficient
of light absorption measurements; and w, the mass per unit area — thickness of the object — is directly
related to the net change in direction of electrons in the object.

In order to use Equation 6.39 to derive the mass of an object transradiated in the transmission electron
microscope, it would be necessary to measure I, point by point, in the electron beam and then sum the
logarithms of the individual measurements. It is simpler to convert I to the photographic density D,
using the photographic emulsion as an analog converter of I. This conversion is linear for density values
below 1.2 and is given by:[465]

$$D = Eb + \delta \qquad (6.40)$$

where E is exposure of the emulsion to the electron beam (I in amperes multiplied by the time); b is a
constant governed by the response of the emulsion to electrons and by the conditions of development;
and σ is the fog produced by the emulsion by diffusely scattered electrons and by the chemical develop-
mental procedure. Linearity for electrons makes it possible to convert electron intensity, Equation 6.39,
directly to photographic density D, Equation 6.40, and to use the relationship of photographic density
to transmission.

$$D = -\log T \qquad (6.41)$$

In considering Equations 6.39 to 6.41, the following relationship was proposed by Bahr:[465]

FIGURE 6.7   An integrating photometer (IPM-2). (Courtesy: Carl Zeiss Instruments, New York.)

$$\left(T - T_o\right)\big/T_\infty \cong \alpha Sw \tag{6.42}$$

where T is the transmission of an image point in the electron micrograph of the object plus its underlying support film, T_o is the corresponding transmission through the supporting film (background), and T_∞ is the transmission of the unexposed photographic material, which is chiefly determined by the fog level σ (Equation 6.40). The transmission difference becomes proportional to the total mass W:

$$\overline{T} - \overline{T}_o\big/T_\infty \cong W \tag{6.43}$$

where $\overline{T}$ is the mean transmission of the entire image area of an object and includes the mean background transmission, $\overline{T}_o$.

It was found experimentally that Equation 6.39, and thus Equation 6.42, are largely independent of the elemental composition of the object. A linear relationship was demonstrated.[464]

6.4.1.1   The Integrating Photometer[466]

The image of the viral particle on the electronmicrograph, plate, or film is analyzed by means of an integrating photometer. One type of integrating photometer, the IPM-2, is shown in Figure 6.7.

The optical pattern is in part analogous to that for Kohler illumination in a light microscope. For measuring transmission in the IPM-2, the measuring aperture is illuminated by the light source through a collecting lens which forms a 1:1 image of the aperture through the condenser in the plane of the electronmicrograph negative. The light passing through the aperture and not deflected by the beam splitter produces a ×5 magnified image of the measuring area on a ground glass screen. Simultaneously, the light source is imaged by the collecting lens into the plan of the selector wheel, and from there via the beam splitter and lens into a photomultiplier. The electrical signal generated is amplified, rectified, and fed to a suitable meter, in this case a galvanometer, and simultaneously to an analog output. The light source is a 50-W photometer lamp, supplied by a separate, precisely stabilized DC power supply. Uniform illumination is achieved by having opal glass in front of the lamp and in front of the photomultiplier.

Ten circular apertures are arranged on a turret so that sizes of 1.5 to 18 mm diameter can be selected and brought individually, precentered, into the beam. In order to compensate for the considerable difference in luminous flux between the smallest and largest apertures, a set of neutral density filters is changed automatically in front of the photomultiplier. Exact compensation is attained by automatic slight modification of the amplification factor. If 100% transmission has been set for unexposed emulsion with one aperture, changing to an aperture of a different size will have little or no effect on this setting.

The integrating procedure may be used with noncircular apertures. For this purpose a second aperture turret is employed into which free-form apertures can be inserted. The IPM-2 is designed so that one can choose to read all values into computer-compatible data storage. Use of the integrating photometer provides a fast, accurate, and convenient method of making mass determinations from electron micrographs. Measurements recorded for computer processing are triggered in rapid sequence with a foot switch. Routinely, 150 to 250 measurements per hour can be made, suggesting that the procedure is not limited by the time required for photographic evaluation, but that electron microscopy and photoprocessing set the pace.

6.4.1.1.1 Measurement of Photometric Transmission: Conversion to Mass. Equation 6.39 would require that every point of the object be measured, followed by transformation of the exponential values and addition to give the total mass. However, no such complications arise when the measured value is related linearly to mass. This fact holds for the transmission over the background to at least a useful approximation. In practice, linearity holds for mass vs. transmission readings for negatives of photographic density ranging from 0.5 to 0.75.

All values of contrast, $T - T_o$, for all image points can be determined with only two measurements, one over the image area, the other in its immediate vicinity. The difference of these two measurements is proportional to the total mass. The transmission T of light through the plate or film is directly related to the electron opacity and hence to the mass thickness w of the object. The more electron opaque the object, the higher the resulting transmission. Thus, the transmission through the photographic plate, T_o, produced by the unaltered electron intensity I_o, which is the background, will always be at the lowest value.

A photometer whose homogeneous light beam of cross section A covers the entire image area of the object measures a mean transmission $\overline{T}$. If the background transmission T_o is homogeneous, the same photometer would measure a mean background transmission $\overline{T}_o$ equal to T_o. Because of the linear relationship between T and w, the difference ΔT of both measurements is a mean mass thickness $\overline{w}$ over the entire area A of the measuring light beam:

$$\frac{\Delta T}{T_\infty} = \frac{\overline{T} - T_o}{T_\infty} = \alpha S w \tag{6.44}$$

S, the scattering cross section (cm²) per gram of substance, is presumed not to vary for different parts of the object. This is especially unlikely for relatively small objects such as virus particles. To the integrating photometer, every object appears to have a mean mass thickness $\overline{w}$ uniformly spread over the area regardless of actual variations in mass thickness, w, or in the shape of the object. The treatment of the proportionality factors and S is discussed below (Section 6.4.1.2).

The apertures of the integrating photometer are chosen to fit the image as tightly as possible so that the contribution of the background T_o to the total transmission measured, $\overline{T}$, will be as small as possible, and the difference, $\overline{T} - T_o$, at a maximum. This condition is illustrated in the case of vaccinia particles in Figure 6.8.[467] The difference in transmission readings is normalized by multiplying it by the area A of the aperture used in the integrating photometer. This product, for the calculation of mass, is referred to as R:

$$R = \frac{\overline{T} - T_o}{T_\infty} A \tag{6.45}$$

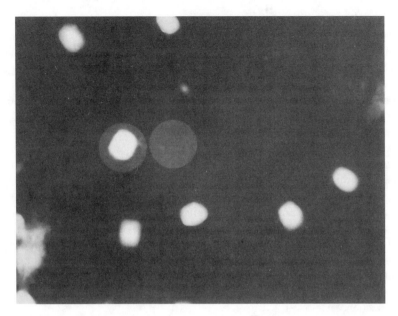

FIGURE 6.8 Vaccinia virions having their integrated transmission ($\overline{T} - T$) values measured. An aperture (left) was selected to cover as closely as possible the area of a vaccinia particle. In this manner, the total transmission, $\overline{T}$, representing the transmission through the virion, T, plus that through the background, T_0, is measured. The background transmission, T_0, is measured by the second aperture (right) which is identical in area to that of the first aperture. The difference of the two transmission readings is ΔT, which when multiplied by the area of the aperture gives the normalized transmission reading (R) for the virion. Original magnification ×60,000. Not shown is the shielded circular area for setting the transmission to 100%, thereby correcting for the effect of photographic fog, T_∞, on transmission. (From Mazzone, H.M. et al., *Methods in Virology*, Vol. 8, Macamorosch, K. and Koprowski, H., Eds., Academic Press, New York, 1984, 103. With permission.)

The value of the photographic fog, T_∞, is present in Equations 6.42 to 6.45. It can be corrected by setting the transmission to 100% in an unexposed area of the electron micrograph, an area expressing only photographic fog. To produce such an area on the negative, a grounded metal disc about 15 mm in diameter is mounted to a suitable part of the microscope camera, as close as possible to the plane of the emulsion, in order to shield a circular area of the negative from exposure. In electron micrographs, the shielded circular area appears transparent on the negative and black on the positive.

6.4.1.1.2   Computer-Aided Operations.   An attractive alternative to integrating photometry with the IPM is scanning-digitizing of the negative. For this purpose a flatbed scanner densitometer, a TV-based digitizer, or a digitizing drum scanner is a prerequisite. In most instruments of this type it is possible to define the area of scan by setting a scan window around the contours of the object image. The window is comparable to the feret diameters of the object projected onto the x and y movements of the scanning instrument. Beyond these mechanical means, there are software alternatives. For example, if the entire negative has been digitized, it can be displayed on a monitor and the object delineated with a light pen or cursor. Or, the object can be delineated by driving the scanning stage so that a fixed point, e.g., the cross hair in an ocular, appears to move around the image of the object, thus indicating to the computer the area to be scanned and reducing the digitizing to the object area only.

An advanced step is the identification of the object area entirely by software through either thresholding or scene segmentation. The first procedural step is to set the scanner on an unexposed portion of the photographic emulsion and adjust transmission to 100%, thus compensating subsequent measurements of photographic fog. Assuming that a digitized image is at hand, a histogram of transmission vs. frequency will assist in separating background from object transmissions, whereupon the sum of the object transmission is calculated as well as the number of object pixels noted.

In order to correct for background the average transmission of one background pixel is calculated from 50 or 100 lowest transmission values. This value is multiplied by the number of object pixels, and the sum is subtracted from the total object transmission. If the object is the image of a polystyrene sphere subjected to electron microscopy at the same magnification and processed under the same conditions as the object, the above procedure gives a transmission value that can be related to an absolute mass value calculated from the diameter of the sphere, corrected for magnification, i.e., the true volume of the sphere multiplied by the specific gravity of polystyrene (1.05 g/cm³). Successive determinations of both total transmission and mass for spheres of different sizes produce a calibration curve of transmission vs. mass, the slope of which may be used in calculations of the absolute object mass.

A distinct advantage of the scanning-densitometric assessment of mass by the contrast method is that object images filling the area of the electron microscope plate can be measured while the integrating photometer is limited to circular areas 2.0 cm in diameter or smaller. Measuring the mass of an entire cell or of cell nuclei can best be done with the scanning approach. A flat object 10 µm in diameter will become in an electron micrograph at ×6000 magnification a disk 6.0 cm in diameter fitting well on a 8.3 × 10.2 cm plate. A calibration sphere 0.756 µm in diameter, however, will be a disk 4.54 mm in diameter. Many nuclei have a size of 50 µm. At a magnification of ×1500 they will be 7.5 cm on the negative. A sphere safely within the range of the penetration of electrons is 0.756 µm in diameter. It will appear to be only 1.13 mm at the same magnification. This size is just at the limit of mass determination by contrast evaluation.

6.4.1.2   Standardization

The procedure for integrating photometric weight determinations must be standardized:

$$W_{unknown} : W_{standard} :: R_{unknown} : R_{standard} \qquad (6.46)$$

where R is defined as in Equation 6.45.

6.4.1.2.1   External Standards.   When using standards that are not electrographed on the same film or plate as the unknown, polystyrene latex (psl) spheres are suitable for mass determination. Because of their shape, the actual mass is determined from their volume and the density (g/cm³) of polystyrene. In attempting to use psl spheres as internal standards, it has been observed that they tend to form aggregates with virus particles. Therefore, it may be necessary to use latex spheres as external standards. When exposed to the electron beam for a brief time the psl spheres do not lose their shape, and they also have an additional advantage in that they can be purchased in an assortment of sizes representing a variety of mass values.

$$W_{psl\ sphere} = \text{volume} \times \text{density} \qquad (6.47)$$

In practice, it is easier to measure the diameter of a sphere d, than to attempt to pinpoint the center of the sphere for the radius measurement. The density of polystyrene is given as 1.05 g/cm³

$$W_{single\ psl\ sphere} = \tfrac{4}{3}\pi \left(\tfrac{1}{2}d\right)^3 \times 1.05 \times 10^{-3}$$

and

$$\tfrac{4}{3}\pi \times \tfrac{1}{8} \times 1.05 = 0.5495 \qquad (6.48)$$

$$W_{single\ psl\ sphere} = 0.5495\ d^3 \times 10^{-3}$$

Since d was measured in millimeters, it is multiplied by 10^{-3}.

Experimentally, a mixture of latex spheres is electrographed separately under the same conditions as the virus of unknown mass. Exposure must be controlled so that the photographic density of the background on the negative is between 0.5 and 0.75, the ideal being 0.60. For experiments kept in this range, transmission of the image of objects, e.g., virus particles is not affected by density differences.

The negatives of the standard and those of the "unknown" must be developed together under standard photographic conditions. For studies reported in this section, the development of electronmicrograph negatives was done in Kodak D 19 solutions diluted 1:2 for 5 minutes, with intermittent agitation every 30 seconds, at a temperature of 21°C.

Each psl sphere is measured for its $\overline{T} - T_o$, and multiplied by the area of the aperture used in the integrating photometer, Equation 6.45, to obtain the value, $R_{psl\ sphere}$. All the R vs. d^3 values for the assorted sizes of psl spheres can be added and averaged or the R vs. d^3 for each sphere size can be plotted. A straight line is obtained whose slope is proportional to αS, the terms defined in Equation 6.39. The average value of R/d^3 or the slope value, is referred to as β in Equation 6.49, which becomes the working equation for the determination of mass W, when using psl spheres:

$$W_{unk} = \frac{0.5495 \times 10^{-3}}{\beta \times M^3} \times R_{unk} \tag{6.49}$$

Note that in Equation 6.49 the magnification M was raised to the third power because the diameter for each psl sphere size was raised to the third power to derive β, which is also present in Equation 6.49. An example of the mass determination of SV40 is given below; psl spheres were used as external standards.

6.4.1.2.2   Internal Standards.   If the standard used in mass determination can be electrographed on the same negative, as would be the case in using a known virus as an internal standard, the proportionality factors of mass, W, and corrected transmission value, R, for the known virus and R for the unknown virus are all that are required. The magnification term does not enter into the calculation:

$$W_{unk} = W_{std} R_{unk} / R_{std} \tag{6.50}$$

After development of the negative, it is analyzed in the integrating photometer. A statistical number of standard virus particles and unknown virus particles are measured for their transmission values ($\Delta T = \overline{T} - T_o$), and ΔT is multiplied by the area of the aperture used in the integrating photometer to obtain the corrected reading, R. By proportionalization, the mass of the unknown virus is calculated from Equation 6.50. An example is given below for the average mass determination of tobacco mosaic virus (TMV) using SV40 as an internal standard.

Logarithmic normal distribution of viral mass — In constructing histograms reflecting the frequencies of corrected photometric readings R, values of the corresponding viral mass exhibit a distinct tendency to be skewed toward the higher side. However, the mass of viruses is distributed normally when considered logarithmically to the base 10. When R is converted to log R and plotted against percentage cumulative frequency, F, in a probability graph, a straight line is obtained, indicating the logarithmic distribution of viral mass, as shown in Figure 6.9. Log normal distributions are common for most biological measurements, including weight, length, height, volume, and/or circumference.[468,469] In plotting log normal distributions in the determination of mass by quantitative electron microscopy, one can observe mass distribution for the mode of the population. One can also observe the deviation in modality, indicating the heterogeneity in the virus population. In an intensive analysis of the mass of vaccina virus particles, Bahr et al.[470] concluded that while genetic control in a presumably homogeneous virus population is strict with respect to quality, the quantity of viral components, presumably other than DNA, varies in a fashion that can be aptly described by a log normal distribution.

Log σ for distribution — As noted above, the mass of viruses is given as the median of the distribution because it is skewed toward higher mass values. In such cases, the more marked the skew, the more

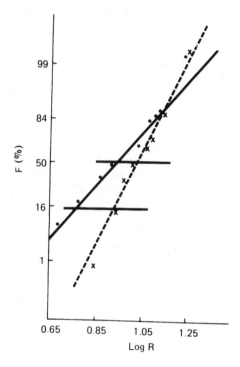

FIGURE 6.9   Log normal distribution of virus particles in a plot of percentage cumulative frequency, F, vs. corrected transmission, log R, in a probability chart. The solid line represents the plot for T2 virus, and the dashed line the plot for *Tipula iridescent* virus. At the 50% point, the value for the average log R can be obtained, and R is used in the calculation of the average mass of the virus population. In addition, log for distribution, the standard deviation, can be obtained as the difference of the readings at 50% and 16%. (From Mazzone, H.M. et al., *Methods in Virology*, Vol. 8, Maramorosch, K. and Koprowski, H., Eds., Academic Press, New York, 1984, 103. With permission.)

misleading is a mean, particularly the standard deviation of the mean, σ. If the distribution is normal, and log R is plotted against F, the percentage cumulative frequency, in a probability chart, the three distributional measures — median, mean, and mode — fall together. The standard deviation, log σ, is identified in each graph between the points of 16% and 50% cumulative frequency (Figure 6.9).

6.4.1.3   Procedure and Treatment of Data to Obtain the Average Mass of Virus Particles

In this section the average mass value for two viruses is obtained using transmission data and employing standards whose mass is known or can be easily calculated. In the first example, an external standard is used, and in the second example, an internal standard.

6.4.1.3.1   *Average Mass of SV40 Using Polystyrene Latex Spheres as External Standards.*[467]   A drop containing virus particles was placed on Parlodion–carbon films (about 50 Å of carbon) and dehydrated through ascending concentrations of acetone to 100% acetone. The sample grids were critical-point dried and placed in an oven at 180°C for 20 minutes to remove the Parlodion, leaving the virus particles supported only by the carbon film. The virus particles were electrographed in a Hitachi HU-12 electron microscope under the following conditions: magnification ×36,800, a 100 µm condenser aperture, a 70 µm objective aperture, 75 kV accelerating voltage. After development and drying, the plates were analyzed in the IPM-2 integrating photometer.

The plate was set emulsion-side up in the photometer. Electrical zero was obtained with the beam blocked by an opaque portion of the stage, and the image of a virus particle was centered on the ground-glass observation screen. An aperture was selected to conveniently cover the size of the image of the virus

TABLE 6.3   Transmission Values for 10 Images of SV40 Particles, Obtained with the Integrating Photometer

Virus Particle	$\overline{T}$	T_o	ΔT	R[a]
1	345	302	43	1.68 ≅ 1.7
2	340	296	44	1.72 ≅ 1.7
3	335	296	39	1.52 ≅ 1.5
4	330	288	42	1.64 ≅ 1.6
5	335	296	39	1.52 ≅ 1.5
6	334	295	39	1.52 ≅ 1.5
7	330	292	38	1.48 ≅ 1.5
8	330	286	44	1.72 ≅ 1.7
9	320	276	44	1.72 ≅ 1.7
10	312	375	37	1.44 ≅ 1.4

Note: Aperture number 2 used.

[a] R = ΔT A. A, the area factor for aperture number 2, is 3.9×10^{-2}.

From Mazzone, H.M. et al., in *Methods in Virology,* Vol. 8, Maramorosch, K. and Koprowski, H., Eds., Academic Press, New York, 1984, 103. With permission.

particle. The carriage was moved so that the bright disk of the aperture image fell clearly in the unexposed area of the plate and the transmission T_∞ was adjusted to 100% on the galvanometer. Next, the image of the virus particle was brought into the aperture, and the total transmission $\overline{T}$ over the image area was measured, i.e., T of virus image plus background T_o. The aperture image was then moved to a free background and T_o was recorded. For 128 SV40 particles, readings of ΔT, representing $\overline{T} - T_o$, were obtained as demonstrated in Table 6.3 for 10 SV40 particles. As noted above ΔT multiplied by the area factor, A, for the aperture used in the integrating photometer, yields R.

The virus particles were then grouped according to their R values. The range of R covered the values observed in measuring the particles. For the 10 SV40 particles noted in Table 6.3, R had a range of 1.4 to 1.7. Assigning the number of particles to R values of 1.4, 1.5, 1.6, and 1.7 would result in 1 for R of 1.4, 4 for R of 1.5, 1 for R of 1.6, and 4 for R of 1.7.

After measurement of the virus particles, the analysis continues as shown in Table 6.4 where n, Σn, and %Σn stand for the number of particles, cumulative number, and percentage of cumulative number, respectively. In order to analyze a logarithmic distribution of the virus particles, log R is assigned to each R value, as shown in Table 6.4. The %Σn is compared to the next higher log R. Thus, %Σn of 2.3 is compared to the log of 1.2, or 0.08. This procedure is followed for the next %Σn in the series, and so on. In order to assign a log R to the last %Σn, i.e., a %Σ of 100 in the series, the range of R was extended by 1 to give 2.0. The log of 2.0 is 0.301, rounded to 0.3, which was used as the log R corresponding to a %Σn of 100.

Log normal distribution of SV40 particles — A plot of %Σn (or F) vs. its corresponding log R is made to obtain the log R at the 50% point, or the average log R. A log normal distribution was obtained for SV40. From the graph, log R at a frequency F of 50% was 0.170. The value of R was 1.479. The next step was to obtain R and w for the polystyrene latex spheres, the standards.

PSL standards — Various sized psl spheres were electrographed separately but under the same conditions as the SV40 particles. The proportionalization is as shown in Equation 6.46.

$$W_{SV40} = W_{psl\ spheres} \times \left(R_{SV40}/R_{psl\ spheres}\right)$$

As noted above, when using psl spheres as standards, the working equation for the average mass determination is Equation 6.49.

TABLE 6.4   Percentage Cumulative Frequency, F,
of Transmission Values, R, for SV40 Virus Particles

R	n	Σn	Fa	Next Higher Log Rb
1.0	0	0	0	0.04
1.1	3	3	2.3	0.08
1.2	11	14	10.9	0.11
1.3	20	34	26.6	0.15
1.4	38	72	56.3	0.18
1.5	22	94	73.4	0.20
1.6	20	114	89.1	0.23
1.7	9	123	96.1	0.26
1.8	2	125	97.7	0.29
1.9	3	128	100	0.30
2.0	—	—	—	—

a %Σn = F.

b Log values are rounded.

From Mazzone, H.M. et al., in *Methods in Virology*,
Vol. 8, Maramorosch, K. and Koprowski, H., Eds., Academic Press, New York, 1984, 103. With permission.

$$W_{SV40} = \frac{0.5495 \times R_{SV40} \times 10^{-3}}{M^3 \times \beta}$$

where M, the magnification, is raised to the third power. The magnification used in the experiment was ×36,800. The term β was obtained from an average of the ratio of R to d^3; d is the diameter raised to the third power of the psl spheres and R is the corresponding normalized transmission value. As noted above, the term β may also be obtained from the slope of the line representing a plot of R vs. d^3 for each psl sphere size. Inserting the corresponding values in Equation 6.49:

$$W_{SV40} = \frac{0.5495 \times \overset{(R)}{1.479} \times 10^{-3}}{\underset{(M^3)}{49.8 \times 10^{12}} \times \underset{(\beta)}{0.3}}$$

$$W_{SV40} = 5.44 \times 10^{-17} \text{ g}$$

The molecular weight (grams/mole) was obtained by multiplying the mass value by Avogadro's number: $5.44 \times 10^{-17} \times 6.02 \times 10^{23}$, or molecular weight for SV40 of 32.7×10^6.

6.4.1.3.2   Average Mass of TMV Using SV40 Particles as Internal Standards.[467]   Electrographing a mixture of SV40 particles and TMV particles and analyzing their images on a negative allows a direct comparison by which the average mass of TMV is obtained according to Equation 6.50.

The average mass of SV40 obtained in the preceding experiment was given as 5.44×10^{-17} g. What is needed is to obtain $\overline{T} - T_o$ transmission values for each of the two viruses, and normalize ΔT with respect to the area of the aperture used in the integrating photometer to obtain R values for SV40 and TMV, respectively, as demonstrated in Table 6.5.

For SV40, a plot of %Σn or F vs. log R gives the 50% value or average log R as 0.85; R is 7.08. In the same manner, a plot of %Σn or F vs. log R gave the 50% value of log R for TMV particles as 1.01; R was 10.23. The average molecular weight of TMV was calculated from Equation 6.50.

TABLE 6.5   Treatment of Transmission Data for SV40
Particles Used as an Internal Standard for TMV Particles

Virus	R	n	Σn	F	Next higher log R[a]
SV40	5	7	7	5.7	0.78
	6	23	30	24.6	0.85
	7	54	84	68.9	0.90
	8	33	117	95.9	0.95
	9	4	121	99.2	1.00
	10	1	122	100	1.04
	11	—	—	—	—
TMV	5	1	1	1.5	0.78
	6	—	—	1.5	0.85
	7	6	7	10.9	0.90
	8	4	11	17.2	0.95
	9	12	23	35.9	1.00
	10	25	48	75.0	1.04
	11	—	—	—	1.08
	12	14	62	96.9	1.11
	13	—	—	—	1.15
	14	2	64	100	1.18
	15	—	—	—	—

[a] Log values are rounded.

From Mazzone, H.M. et al., in *Methods in Virology*, Vol. 8, Maramorosch, K. and Koprowski, H., Eds., Academic Press, New York, 1984, 103. With permission.

$$MW_{TMW} = \left(MW_{SV40} \times R_{TMV} / R_{SV40} \right)$$

$$= \left(32.7 \times 10^6 \times 10.23 \right) / 7.08$$

$$= 45.8 \times 10^6 \, g/mol$$

It should be noted that when internal standards are used in the determination of mass, the magnification and β term are not considered.

Of the two procedures for determining mass, the use of external standards or the use of internal standards, the virologist would be tempted to choose the latter. Such a choice is permitted provided that in the mixing of the two viruses, there is not extensive aggregation to make transmission measurements difficult, nor are any physical or chemical alterations noted for the virus particles. If any of these conditions is observed, the researcher can use the method of external standards to obtain the average mass value of the virus particles. Consider also that psl spheres store better than most viruses. Therefore, if there is any doubt concerning the structural integrity of the virus particles to be used as the internal standard, it would be prudent to use psl spheres, and the external standard procedure for determination of the virus particles of unknown mass.

6.4.1.4   High Voltage Quantitative Transmission Electron Microscopy

In obtaining mass values of viruses, procedures utilizing either the transmission electron microscope (TEM) or scanning transmission electron microscope present no problem in obtaining images for analysis. However, some viruses, e.g., some insect viruses, have an intimate relationship with inclusion bodies in that virus particles are located within such structures (see Figures 3.14 and 3.16). In TEM studies, obtaining an image of inclusion bodies for mass determination is not possible with conventional electron microscopes. The inclusion body is too thick to give an image that can be analyzed for mass determination.

Early experimentation in electron optics in the 1940s indicated that increases in accelerating voltages in electron microscopy improved specimen penetration and resolution. In Europe and the U.S., a number of investigators experimented with accelerating voltages above 100 kV, the common voltage level in conventional electron microscopy. During this period van Dorsten et al. described a 400 kV electron microscope and electrographed yeast cells at 350 kV. This was the first report of high voltage electron microscopy of biological specimens.[471]

In the 1950s high voltage electron microscopes (HVEM) began to be constructed in Japan, England, and Russia. In the 1960s, a microscope operating at 750 kV was built under the leadership of Dr. V.E. Cosslett at the Cavendish Laboratory.[472] During this period Dupouy and coworkers at the Laboratoire d'Optique Electronique du Cente National de la Recherche Scientifique at Toulose built the world's first 1 million-volt electron microscope.[473] Toulose was established as the birthplace of megavolt electron microscopy with the construction of a 3 million-volt microscope in the 1960s and 1970s.[474] By the 1980s, HVEMs were operating in a number of countries. In the U.S., the National Electron Microscopy Center was established at the Lawrence Berkeley Laboratories (LBL) of the University of California. Here, HVEMs were set into operation.

Instrumental details of the HVEM are presented in reviews by Dupouy[475] and by Cosslett.[472] Biological aspects are considered in reviews by Hama[476] and by Glauert.[477] The utilization of the HVEM in virology was discussed by Mazzone et al.[478]

6.4.1.4.1   Obtaining Mass Values of Insect Viral Inclusion Bodies.   As a class of insect viruses the Baculoviruses are natural enemies of many pest insects and are used to control their populations.[479] Baculoviruses contain DNA and exist in nature as rod-like (*baculo* = rod) entities contained within a proteinaceous structure, the inclusion body. Generally, these inclusion bodies are of two shapes: polyhedral and capsular.

The polyhedral type of baculovirus multiplies in the nuclei of insect cells, and is referred to as a nucleopolyhedrosis virus (NPV). The NPVs have many viruses contained within each inclusion body. The capsule (granule) type of inclusion body may multiply in the nucleus or cytoplasm of insect cells, and is referred to as granulosis virus (GV). The GVs have only one virus per capsule, and rarely two.

The Baculoviruses used as insecticides in the field can be grown in considerable quantity and purified in high yields by zonal rotor purification.[153,480] Following purification an average mass value for several NPVs was obtained (Table 6.2). The same procedure for measuring mass values of inclusion bodies with the HVEM was followed as described for the procedure using conventional electron microscopes. The high penetrating power of the HVEM produced images of inclusion bodies which could then be analyzed for their mass values.[478]

6.4.2   Counting of Particles to Obtain the Mean Molecular Weight

In 1949 Williams and Backus[481] demonstrated that it is possible to obtain molecular weights of viruses by direct particle counting with the transmission electron microscope. A very small drop of known volume of virus sample is placed on a coated grid. Prior to applying the drop, a known amount of polystyrene latex particles is mixed with the unknown virus sample. When the limits of the original drop are found on the grid, the number of particles appearing in the drop pattern is counted to give the number of particles per unit volume. This figure, together with the mass per unit volume obtained by weighing the dried residue from a measured volume, is then used to calculate a value for the mean molecular weight of the particles. From a knowledge of the concentration of latex particles in grams per milliliter, the number of latex particles per unit volume in suspension is readily calculated. To find the volume of a drop, it is then necessary only to count the number of latex particles appearing in the drop pattern.

No measurement of dimensions of the particle is needed. All that is required is a count of the virus particles and latex spheres in the drop. With the counting procedure, Williams and Backus calculated the molecular weight of bushy stunt virus as 9.4 million.

Polson et al.[482] used hemocyanin as a standard to determine the molecular weights of TMV and the insect virus of *Nudaurelia cytherea capensis* by the counting procedure. The calculated molecular weights were in good agreement with values obtained by other investigators (Table 6.2).

6.4.3   Quantitative Scanning Transmission Electron Microscopy

The high resolution scanning transmission electron microscope (STEM) became a research instrument in the late 1960s as a result of development of a practical field emission electron gun at the University of Chicago by Crewe and coworkers. The high brightness of this source made possible diffraction-limited probe sizes and resolution equivalent to the best conventional electron microscopes. Contrast was much higher as a result of efficient dark field imaging with an annular detector.[483]

The STEM has been used to determine the mass of objects, including viruses.[484,485] There is a direct relationship between specimen mass thickness and the annular detector signal (or energy loss signal) in the STEM. The number of electrons striking the annular detector, n_s, is given by Beer's law:

$$n_s = DA\left(1 - e^{-N\sigma/A}\right) \cong DN\sigma \tag{6.51}$$

where D is the dose (el/Å^2) incident on a picture element of area A containing N atoms of a certain type, and σ (Å^2) is the cross section for an atom in A to scatter an electron from the conical illuminating beam onto the annular detector.

The treatment of Engel is followed in this discussion.[486] Elastic as well as inelastic scattering has been investigated both theoretically and experimentally using scanning transmission electron microscopy. For a thin specimen, the number of electrons elastically scattered, N_e, by proteinaceous matter can be presented by:

$$N_e = \langle\sigma_e\rangle nN_0/A \tag{6.52}$$

where $\langle\sigma_e\rangle$ = the average elastic scattering cross section of atoms that constitute the specimen, n = the total number of atoms within the irradiated area A, N_0 = the total number of electrons incident on area A. The single scattering approximation is satisfied if $n\langle\sigma_e\rangle \ll A$. The fraction of electrons collected by the annular detector is evaluated from the elastic scattering amplitude f_{ce} of carbon and the detector geometry:

$$\epsilon(\beta) = 2\pi \int_\beta^{\pi/2} \left|f_{ce}\right|^2 \sin\theta\, d\theta / \sigma_{ce} \tag{6.53}$$

ϵ is the collection efficiency of the annular detector vs. the half angle β, θ = the scattering angle, σ_{ce} = elastic scattering cross section of carbon, f_{ce} can be calculated from models for the atomic scattering potential or evaluated from literature tables, ϵ (β) is computed from tables for keV electrons.

According to Equations 6.52 and 6.53 the mass P of a proteinaceous specimen is given by:

$$P = n\langle M\rangle = N_{AD}\langle M\rangle A/N_0\epsilon\langle\sigma_e\rangle \tag{6.54}$$

where $\langle M\rangle$ = the average mass of atoms that constitute a protein, $N_{AD} = N_{0e}$, the number of electrons collected by the annular detector.

Since the specimen is carried by a supporting film the contribution of the corresponding background has to be determined and subtracted from N_{AD}. In practice, the area A that comprises the particles to be measured is outlined by the investigator, the average background of adjacent areas is evaluated, normalized to A and subtracted. It is convenient to normalize N_{AD} to the dose $D = N_0/A$ and to establish the calibration factor C. The calibration factor relates the number of scattered electrons to the mass of the virus and is experimentally determined from a standard:

$$C = \langle M \rangle / \epsilon \langle \sigma_e \rangle \tag{6.55}$$

The mass of the specimen is then given by:

$$P = C\left(N_{AD} - B\right)/D, \tag{6.56}$$

where B = background due to supporting film scattering of the area A.

As with quantitative TEM, particles are measured in the STEM, one at a time, so that the spread in size as well as the average can be determined accurately. In determining the mass of viruses by QSTEM, an external or internal standard may be used. As for QTEM, the external standard is generally polystyrene latex spheres, and the internal standard a virus or protein. Many QSTEM studies employ TMV, an internal standard.

Particles to be measured are selected from images displayed on a television monitor. By direct connection of the STEM to a computer, and utilizing an image processor and related software,[487,488] rapid densitometry is achieved. The photographic process is omitted. After background subtraction, subimages are integrated to yield measurements of mass.

The possibility of mass loss by electron irradiation of the specimen can be controlled. Freeman and Leonard[434] reported that after a typical dose of 100 electrons/nm^2, mass loss was of the order 1 to 2%. Results obtained using a cold stage at 50 K showed that the mass loss was reduced to less than 1% at this dose. The loss rates for different particles were very similar, with the mass ratio remaining constant up to a total dose of 2000 electrons/nm^2. Wall and Hainfeld[489] reported that mass loss is a linear function of dose, electrons/Å^2, and is roughly 2.5% for protein at 10 electrons/Å^2 and -150°C.[489] TMV, used as an internal standard, loses mass at the same rate so, to first order, the effect of mass loss can be corrected by scaling with TMV. At high dose, 100 to 1000 electrons per Å^2, most specimens reach a "stable" state with no further mass loss.

Mass measurements by QSTEM may be done on the whole object or any part of it, or in terms of per unit length of the specimen. Mass values for some viruses are presented in Table 6.2.

7

Hydrodynamic Sizing
and Solvation

The size of a virus is one parameter of value in its characterization. Electron microscopy is conveniently used for sizing viruses in their anhydrous state. Other methods such as laser light scattering spectroscopy (dynamic light scattering), high-flux X-ray and neutron scattering, and the resistive pulse technique, are also available and show advantages in precision, resolution, and rapidity of measurements of macromolecules in solution. High-resolution sizing analysis of viruses by the resistive pulse technique may also reveal structural variants that could not be observed readily by dynamic light scattering and electron microscopy.[504]

As noted in Chapter 5, laser light scattering is widely used to measure the diffusion constant and hydrodynamic measurements of biosolutes. It is well suited for measurements on pure monodisperse samples, but is less suitable for the analysis of polydisperse samples. The resistive-pulse technique, however, can be readily used to size components in a complex mixture. It may be particularly useful for studies in which polydispersity is naturally occurring, and in which purification into single components is difficult to achieve.[504] High-flux X-ray and neutron scattering provide a very important feature of solution scattering. These techniques offer a multiparameter characterization of the overall structure of a macromolecule under physiological conditions.

7.1   Laser Light Scattering[262]

The theory of laser light scattering was discussed in Chapter 5, Section 5.1.8.2. The discussion presented here will be extended to include measurements by which the parameters of size, e.g., radius/diameter and water of hydration can be ascertained. In laser light scattering either the power spectrum or its Fourier transform, the autocorrelation function, of the scattered light intensity fluctuations can be measured to yield information about the diffusion constant of the scattering particles.

For a set of homogeneous particles the scattered light intensity autocorrelation function is $e^{-2DK^2 \gamma}$, where γ is the time delay, D is the diffusion constant, and K is the optical scattering vector. Thus, light intensity fluctuations decay away exponentially with a time constant of $(2DK^2)^{-1}$. This decay can be measured precisely with the autocorrelation computer. By use of the Strokes-Einstein equation,

$$D = kT / 6\pi \, \eta r \qquad (7.1)$$

the hydrodynamic size of the particles can be obtained. In this equation k is Boltzmann's constant, η is viscosity, and r is the radius of the equivalent hydrodynamic sphere. If the particles are not homogeneous in size, a complication arises in that the autocorrelation function becomes a sum of exponentials. One can obtain some measure of the heterogeneity by noting to what degree the logarithm of the autocorrelation function departs from a straight line when plotted against time delay. However, for small-scale polydispersity involving several multiplets, the curvature may be smaller than the scatter in the data owing to a finite signal-to-noise ratio. The spectrometer is usually calibrated by measuring the hydrodynamic size of standard particles of polystyrene latex spheres (Dow Chemical Co., Indianapolis, Indiana).[504]

Utilizing dynamic light scattering the translational diffusion coefficient and the effective hydrodynamic or "Stokes" radius can be obtained relatively rapidly. These measurements can take less than one minute in some cases. Dynamic light scattering is especially useful for looking at the time-course changes in size of assembling/disassembling systems, e.g., the kinetics of head-tail associations of T-type bacteriophages,[505] or the effect of removal of calcium ions on the swelling of Southern bean mosaic virus.[270] In the dynamic light scattering technique, commercial software, e.g., CONTIN is available for inverting the autocorrelation data directly to give distributions of diffusion coefficient and equivalently particle size.[271,273]

7.2   High-Flux X-Ray and Neutron Solution Scattering[412]

High-flux X-ray and neutron scattering techniques, noted in Chapter 6, are used to study the overall structure of biological macromolecules in the solution state. Scattering can be used to monitor conformational changes. The scattering analyses can be quantitatively compared with other physical data obtained from electron microscopy, sedimentation or diffusion coefficients, or crystallography. Analysis of the scattering curve, I(Q), measured over a range of Q, leads to the molecular weight and the radius of gyration. It also provides the maximum dimension of the macromolecule.

Perkins[412] describes how modeling of scattering curves extends the interpretation of the scattering analyses. In this respect the use of small spheres to model the scattering curves is the most flexible and powerful method. Other approaches to calculating scattering curves are based on spherical multishell models or the use of prolate or oblate ellipsoids. The advantage of modeling is that it can rule out structures that are incompatible with the scattering curves.

In sphere modeling the macromolecule, e.g., a protein, is subdivided into cubes, which are represented by the same number of overlapping spheres of the same volume. The total volume of the spheres is set equal to the dry volume of the protein as observed by crystallography[389,506] for neutron data, and to the dry volume plus the volume of the hydration shell surrounding the protein for X-ray data. A hydration of 0.3 g water per gram of protein is usually assumed, and the volume of a protein-surface-bound water molecule is 0.0245 nm^3 in place of the free water volume of 0.0299 nm^3.[389] These procedures have been tested for both X-ray and neutron scattering using a known crystallographic structure.[507] In the absence of a crystal structure, the sphere modeling procedure determines the simplest triaxial object that will account for the scattering curve, starting with the value of the radius of gyration determined from the curve at lowest Q.[412]

A range of models can be tested by a trial and error process against the experimental data in order to identify a number of possible structures. The resulting model offers a curve that is equivalent to that seen experimentally. The scattering model can be compared with the modeling of frictional coefficients based on hydrodynamic spheres.[507-511] This feature acts as a further control of the modeling, since it is a fully independent physical measurement. In this case, the radius of gyration from scattering is checked for its compatibility with experimental sedimentation or diffusion coefficients.[412]

7.3   Resistive-Pulse Technique[512]

This technique has as its basis the theory underlying the Coulter Counter apparatus (Coulter Electronics, Inc., Hialeah, Florida). In this instrument the particles in an electrolyte solution are caused to flow, by pressure differences, through a small current-carrying aperture. The momentary changes of voltage that occur as the particles pass one by one through the aperture are counted and classified according to magnitude. The small pulses arise from momentary displacement of the electrolyte by the relatively insulating particles and are closely proportional to the volume of each particle. The constant of proportionality depends to some extent on the shape, aspect, and size of the particle. In refining the technique for greater sensitivity, Nucleopore filter material was developed at the General Electric Research and Development Center, Schenectady, New York. Uniform pores of any diameter from 0.010 µm on up may be etched through nonconducting materials, such as Lexan polycarbonated plastic sheets, where high energy nuclear particles have left damaged tracks. Individual pores of submicrometer diameter were isolated and used as apertures to analyze particles of various macromolecular size. This development resulted in the Nanopar analyzer.

In resistive-pulse analysis, spherical particles are measured with respect to standard latex spheres of diameter, d, through the equation:

$$d = d_0 \left(\frac{\Delta E}{\Delta E_0} \right)^{1/3} \left[1 + 0.8 \left(\frac{d_0}{\mathscr{D}} \right)^3 \left(\frac{\Delta E}{\Delta E_0} - 1 \right) \right]^{-1/3} \tag{7.2}$$

where ΔE and ΔE_0 are the voltage-pulse amplitudes produced by particles of diameters d and d_0, respectively, as they pass through the pore, and $\mathscr{D}$ is the diameter of the pore.

If the particle is not a sphere, but may be approximated as an ellipsoid of revolution, then its volume V_e may be readily determined in terms of the resistively equivalent diameter of Equation 7.2 and a form factor, f_e, namely

$$V_e = \pi d^3 / 4 f_e \tag{7.3}$$

The form factor f_e is given by

$$f_e = \frac{1}{1 + n_\perp} + \left(\frac{1}{1 - n_\parallel} - \frac{1}{1 - n_\perp} \right) \cos^2 \beta \tag{7.4}$$

where $n_\perp$ and $n_\parallel$ are extensively tabulated demagnetization factors for a field applied perpendicular and parallel, respectively, to the axis of revolution, and β is the angle between the axis of revolution and the field. It is observed experimentally that nonspheroidal particles generally pass through a pore with the long axis parallel to the axis of the pore, so that f_e varies from 1.5 for a sphere to 1 for a long rod. The approximate shapes in a polydispersed sample are obtained through electron microscopy.[504]

7.4 Hydrodynamic Radii and Solvation

For a macromolecule the dry radius is defined as:[282]

$$R_d = \left(3 M \overline{v} / 4 \pi \, N \right)^{1/3} \tag{7.5}$$

where $\overline{v}$ is the partial specific volume of the macromolecule and N is Avogadro's number. R_d can be regarded as that of an equivalent unsolvated sphere. The radius of a hydrodynamically equivalent sphere, R_h, can be defined using the Stokes-Einstein equation:

$$R_h = kT / 6 \pi \, \eta \, D^0 \tag{7.6}$$

where k is the Boltzmann constant, η is the viscosity of the solvent, T is the absolute temperature, and D^0 the zero concentration limit of D.

The ratio of these radii is a function of both the solvation and the degree by which the shape deviates from a sphere:[513]

$$R_h / R_d = \left(f/f_0 \right) \left(\left[\bar{v} + \delta \right] / \bar{v} \right)^{1/3} \tag{7.7}$$

where f/f_0 represents the effect of shape on the frictional coefficient, the second term in the parentheses is the ratio of the solvated to dry volume, and δ is the volume in cm³ of solvent associated with each gram of dry macromolecule.

For $f/f_0 = 1$, the fraction of the virus hydrodynamic volume occupied by solvate is given by:[282]

$$F \equiv 1 - \left(R_d / R_h \right)^3 \tag{7.8}$$

The volume percentage of water can also be estimated using the molecular weight value and the equation:[306]

$$r_i = \frac{3M_i}{4\pi A} \left(\bar{v}_{2i} + \delta_i \, v_t^0 \right)^{1/3} \tag{7.9}$$

where r_i = the radius of the hydrated particle, M_i and $\bar{v}_{2i}$ are, respectively, the dry mass and partial specific volume of the *i*th particle, A is Avogadro's number, and v_i^0 is the partial specific volume of the free solvent, $V_{2i} = (V_{2i} + \delta_i V_i - \delta_i v_i^0)$, where v_{2i} and V_i are, respectively, the partial specific volumes of the dry particle and the bound solvent; δ_i may be interpreted as measuring either deviations from sphericity or degree of solvation.

The volume percentage of water = $1 - M/A_\rho/(4\pi r^3)/3$, where M is the molecular weight of the virus, ρ is the particle density (the reciprocal of the partial specific volume), and r is the hydrodynamic radius as determined from the diffusion coefficient.

Particle volumes may be determined from the hydrodynamic diameters.[314] The volume occupied by the macromolecular components together with specifically bound water is M/A_ρ where M, A, and ρ have their usual significance.

Tables 7.1 and 7.2 show, respectively, hydrodynamic data of sizing and solvation for several viruses by laser light scattering and the resistive pulse technique.

TABLE 7.1   Hydrodynamic Properties of Viruses Determined by Laser Light Scattering

Virus	Diameter (nm)		Water of Hydration cm³ Water/g Dry Virus	% Water by Volume
	Dry	Hydrodynamic		
Phage N4[325]			1.53	
Phage R17[282]	20.6	28	1.02	60
Phage QB[282]	21.2	30.2	1.22	65
Phage PM2[282]	49	66	1.1	59
Phage T7[282]	47	66.6	1.8	65
Tomato bushy stunt[282]	27	34.4	0.75	52
Turnip yellow mosaic[293]	27	30.4		
Avian myeloblastosis[314]	134	142		57
Rauscher murine leukemia[314]		150		69
Murine mammary[306]		144		71
Rous sarcoma[306]		147		64
Feline leukemia[306]		168		69

TABLE 7.2 Hydrodynamic Sizing of Viruses Determined by the Resistive Pulse Technique

Virus	Hydrodynamic Diameter (equivalent sphere) (nm)	Volume $\times 10^{-15}$ cm^3
Rauscher murine leukemia[512]	122.3	
Simian sarcoma[512]	109.7	
Mason-Pfizer monkey[512]	140	
Cat virus RD 114[512]	115	
Feline leukemia[512]	127.4	
Phage T2[512]	96	0.51
Gypsy moth[504]		
Unenveloped rod	106	0.62
Enveloped rod	160	2.15
European pine sawfly[504]		
Unenveloped rod	97	0.48
Enveloped rod	125	1.02
Tipula iridescent[504]	180.5	3.07

8

Available Resources

8.1 Instrument Use

A major source for making instrumentation available is the National Center for Research Resources (NCRR) of the National Institutes of Health. NCRR was established in 1990 with the merger of the Division of Research Resources (DRR) and Division of Research Services (DRS). Since the early 1960s, DRR had provided extramural research resources to investigators supported by the National Institutes of Health and DRS provided resources to NIH intramural researchers.

The Shared Instrumentation Grant Program of the NCRR helps researchers meet critical instrumentation needs by:

- Affording investigators cost-effective access to essential, technologically sophisticated equipment that would be prohibitively expensive to support on a single grant application
- Maximizing the utility of federal research funds by allowing a number of scientists with similar large instrumentation needs to share such equipment
- Facilitating informal, yet essential, interactions among scientists, thereby catalyzing mutually rewarding new research collaborations

Another means of assuring that researchers have access to resources is through focal points of concentrated resources within a research establishment. Biomedical Technology Resource Centers, sponsored by NCRR, accomplish this by enabling access to state-of-the-art technology and advanced equipment that is among the best in the world. Investigators can take advantage of available technologies even if they are not directly affiliated with these centers. The over 60 designated resource centers throughout the United States fulfill the following functions:

- Enable broad access to state-of-the-art technology and instrumentation
- Promote cost-effectiveness through sharing
- Foster the development of new technologies and novel applications of existing technologies, including the computers and software on which research has increasingly grown to depend
- Further communication within the biomedical research community through publications, seminars, workshops, and conferences

The National Center for Research Resources publishes a directory to promote access to a network of biomedical technology resource centers it supports throughout the nation. NCRR encourages qualified biomedical scientists to take advantage of the unique capabilities of the centers listed in this directory, and welcomes comments on their usefulness and the need for additional technologies.

8.1.1 Sedimentation Experiments

The National Analytical Ultracentrifugation Facility of the Biotechnology Center at the University of Connecticut, Storrs, Connecticut, was established in 1988 by a grant from the National Science Foundation. It houses Beckman analytical ultracentrifuges with their associated electronic equipment, computers, and scientists and engineers. The facility is available to investigators for basic research, and conducts training workshops on theory and use of the analytical ultracentrifuge.

Beckman Instruments of Palo Alto, California, offers training courses in sedimentation analyses at the Beckman Centrifuge Development Center. Conducted by Beckman scientists and technicians, training courses cover basic instrument operation, cell assembly, experimental design, data analysis, and trouble-shooting/routine maintenance.

Beckman Instruments publishes a newsletter, *Exploration,* designed to disseminate applications and technique information on sedimentation analyses. The newsletter also updates user group meetings and symposia, national and international.

Consultation and training in sedimentation field flow fractionation is also possible. The point of contact is the Field-Flow Fractionation Research Center, Department of Chemistry, University of Utah, Salt Lake City, Utah 84112.

8.1.2   Scattering Studies

At the National Institute of Standards and Technology (NIST) in Gaithersburg, Maryland, training and information are provided on light scattering, small angle X-ray scattering, and small angle neutron scattering. Concerning the last methodology, a second 30 meter small angle neutron scattering (SANS) instrument has gone into operation at the Cold Neutron Research Facility (CNRF). This instrument will serve as part of an endeavor to establish a Center for High-Resolution Neutron Scattering. The SANS spectrometers are capable of observing nanoscale virus structures. Under sponsorship of the National Science Foundation, a large portion of the available research time on the new spectrometer is available to the research community. It is proposed that the Cold Neutron Research Facility will include 15 state-of-the-art experimental stations available to qualified researchers. (Source: *Chemical & Engineering News,* September 21, 1992.)

Training in light scattering is also possible through contact with commercial companies, e.g., Wyatt Technology (Santa Barbara, California) and Particle Sizing Systems (Santa Barbara, California).

8.1.3   National Resources in Electron Microscopy

A number of scanning transmission electron microscopes are available for use as part of the National Center for Research Resources program. These facilities include the following: the Biological Scanning Transmission Electron Microscopy Facility at the Brookhaven National Laboratory, the scanning electron microscope facility as part of the Integrated Microscopy Resource at the University of Wisconsin at Madison, the Facility for High Resolution Electron Microscopy, Arizona State University at Tempe, and the Nanofabrication Facility, Cornell University at Ithaca.

High voltage electron microscopes are also available through the NCRR program. These include the HVEM facility at the University of Colorado at Boulder, the HVEM facility at the University of Wisconsin at Madison, the HVEM facility of the New York State Department of Health at Albany, and the electron microscope facility of the Lawrence Berkeley Laboratory of the University of California at Berkeley. The latter resource has been designated the National Center for Electron Microscopy (Figure 8.1).

8.1.3.1   Remote Operation of Electron Microscopes

The High Temperature Materials Laboratory (HTML), a part of the Oak Ridge National Laboratory, has as a primary goal the remote operation of transmission and scanning electron microscopes. In this regard, the objective is that a remote user should be able to work with these instruments as if physically present at the computer controlling the microscope.[514]

The HTML is an all digital laboratory in that all of the electron microscopes are equipped for high quality digital imaging. The use of film and a darkroom have been abandoned completely. Eliminating film resulted in an immediate increase of data throughput on the computer network. Macintosh computers are connected either via Ethernet at 10 Mbps or switched Fiber Distributed Data Interface (FDDI) at 100 mbps. The server is NT-Windows based and 24 GB of disk space (Raid Level 5) are continuously

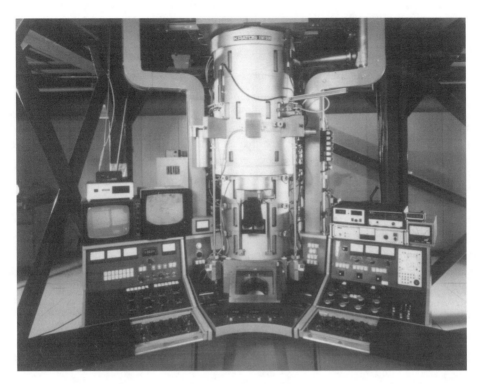

FIGURE 8.1   The research work area of the high voltage transmission electron microscope, at the National Center for Electron Microscopy. (Courtesy: Ernest Orlando Lawrence Berkeley National Laboratory, University of California.)

present on the network as data and image storage. After 6 months, or when memory becomes low, the data are permanently moved onto CDs.

To make remote microscopy more universally available required the use of commercial software only and demanded scripting capabilities. Such requirements were met by: (1) expanding the de facto standard software Digital Monograph (a standard application for the Macintosh platform) to include a complete set of SET and Get commands for microscope control, (2) using the software Timbuktu Pro (Farallon Computing Inc., Alameda, California 94501-1010) to provide remote control capabilities, and (3) using CUSeeMe (available via file transfer protocol [FTP] from Cornell University) and other video conferencing hardware and software for telecommunication.

For remote control, the HTML presently uses the following conventions for both TEMs and SEMs. The "m", "s", and "f" keys on the (local or remote) keyboard activate the controls for magnification, stage, and focus controls, respectively. The arrow keys up/down in/decrease the active parameter, while the left/right arrow keys de/increase the step size for each parameter. Moving the stage deviates from this convention, as movements in all four directions are required. The step size is changed in this case by using the shift or option key in combination with the left/right arrow keys.

As the available bandwidth is limited, the displayed images are usually equal to or less than 256 × 256 pixels. Full size images, lk × lk, can be obtained by using the tabular key of the keyboard. Once the image is acquired and saved, it can be downloaded by the drag-and drop feature of TimbuktuPro.

For remote testing and use of a TEM or SEM at Oak Ridge National Laboratory, HTML asks that a request be sent by e-mail to: "remote EM@ornl.gov".

A highly successful application of non-world wide web networking technology is the multiuser–SEM environment developed at Iowa State University.[515] This facility links a series of Macintosh workstations to the SEM over a local area network, so that users at any workstation can have complete control of the SEM, subject to limitations on the number and variety of operating parameters accessible via the computer interface. The Iowa system uses a custom program to control the microscope, and the network package "Timbuktu" to provide network connectivity and control. Although the laboratory has primarily been developed to support SEM teaching within the university, CUSeeMe has been utilized to provide video conferencing facilities between the SEM operator and observers at remote locations.

8.1.3.2   World Wide Web-Controlled Scanning Electron Microscope Availability

Chand et al.[516] describe a system for remote control of a scanning electron microscope over the Internet using the world wide web. As is well known the world wide web is the largest online collection of documents in the world, with institutions, both academic and commercial, and private individuals contributing to this dynamic, distributed information base. Using WWW browsers is now the most popular and most effective means of accessing online material, whether it be plain text, images, and audio or visual clips. In the paper by Chand et al., there are several major applications rendered feasible by the remote control system, namely, remote control, remote diagnosis, virtual microscopy for education, and real-time collaboration by instrument sharing.

A software system was designed and implemented for remote control of a state-of-the-art research grade SEM, the LEO 440. The instrument is fully PC-based and runs under Microsoft Windows with all SEM parameters controlled and interrogated by the LEO control software. In addition, an Application Programmers Interface and a Network Dynamic Data Exchange (NetDDE) interface allow other Windows-based application software to communicate with the SEM. The NetDDE interface provides a natural safe entry point for other software and forms the link between the SEM and the remote control system. The system configuration developed can be logically divided into three sets of components differentiated geographically: the target site, where the PC-SEM is located, the remote site, where the operator is located, and the intervening network (LAN or Internet).

The advantages of the remote control system as developed by the authors are

- Platform independence at the remote site and virtual platform independence at the target site
- Relative ease of adapting user interfaces based on hypertext for different applications
- Rapid software development by utilizing existing generic tools and packages
- Minimization of development cost, again through the use of generic tools
- Physical layer of networking, whether modem, ethernet, or optical fiber, that is transparent to the system and, with the exception of bandwidth limitations, transparent to the user
- Networking agents providing encryption facilities and password protection for secure communication channels

Future work will involve developing Java programs to handle all aspects of instrument monitoring, as well as image processing and analysis. It is also intended to adapt CUSeeMe so that it becomes a "plug-in" to Netscape Navigator, thus removing the need to have both Netscpe and CUSeeMe windows competing for limited desktop area. Usability studies are to be conducted into user interface design for various remote applications.

8.2   Computer Programs

Computers form an integral part of the methodologies discussed in this book. Numerous computer programs are written to analyze the data. Two of these programs are discussed below.

Currently, the Optima series of analytical ultracentrifuges are shipped with computer programs such as ORIGIN Scientific Graphics and Data Analyses in Windows and XL-A Data Analyses, v. 2.01. ORIGIN provides the foundation of curve fitting algorithms, display flexibility, and ultimate customization upon which user application can be built. XL-A Data Analysis is a package of such applications that helps the novice and experienced users alike apply accepted analysis protocols for determining association constants and stoichiometries, $M_{W,APP}$, s, D, and other thermodynamic and hydrodynamic properties of macromolecules.[517]

Data from experiments involving the Optima series of analytical ultracentrifuges can be written in software language compatible with ORIGIN. UltraScan 2.5 is a Windows/ORIGIN-integrated software package for acquistion and analysis of sedimentation velocity and sedimentation equilibrium data acquired with the XL-A analytical ultracentrifuge.[518] The following functions are supported:

Data acquisition: a windows interface program to the Beckman Data Acquisition program XLA.EXE. allows real-time data display during data acquisition — one observes data on the computer screen at the same time as it is being collected by the XLA.

Sedimentation velocity: Global data editing routine, standard- and high-resolution or boundary zooming G(s) data analysis, using the method of van Holde-Weischet, second moment analysis, data smoothing, finite element method for creating synthetic velocity data specifying $s_{20,w}$, $D_{20,w}$ and concentration, internal buffer tabulations for commonly used buffers to correct for hydrodynamic parameters, protein $\bar{v}$ calculation based on amino acid sequence.

Sedimentation equilibrium: data editing routine with either automatic baseline estimation or subtraction of the baseline obtained from overspeeding the centrifuge at the end of the run. The results are displayed graphcially and data can be displayed as $\ln(C)$ vs. r^2, M_w vs. C, M_w vs. r^2, and as integration plots.

Light scattering: a number of computer software programs are available from which the translational diffusion coefficient, molecular weight, and other measurements are obtained. For molecular weight software, Wyatt Technology, a leading manufacturer of light scattering equipment, offers AURORA. This program performs almost instantaneous absolute molecular weight determinations and draws Zimm plots on-screen using data collected and processed by Wyatt's Dawñ data collection programs. AURORA calculates absolute weight-average molecular weights, root mean square sizes, and second virial coefficients. Linear, quadratic, cubic, and higher-order polynomials may be used for extrapolating the Zimm plot curves to handle data from macromolecules.

Table 8.1 lists a number of computer programs used in sedimentation experiments, scattering studies, and crystallographic data treatment, including virus crystallography.

8.3 Databases

8.3.1 A Database for Viruses

In 1995, a subcommittee of the International Committee on Taxonomy of Viruses (ICTV) met in Doorn, The Netherlands, to begin developing a universal virus database. This public-domain database will be operational through a computer network as well as by CD-ROM where necessary. It will be accessible to scientists and clinicians around the globe.

The international group of virologists planning this project expects to assemble a diverse range of virus data, including information about molecular, pathologic, and epidemiologic properties of most, if not all, known viruses. In addition, the database will be illustrated with morphological and cytopathologic data, including electron micrographs and crystallographic structures.

A prototype plant virus database has been reported.[528] A group of virologists and computer experts at the Australian National University and at London University have compiled a database of almost 1000 plant viruses with hundreds of descriptors. This database has been used to compile books for use in the field by agricultural and other workers trying to identify plant virus infections from observable symptoms.

This unwieldy taxonomy is part of what makes a comprehensive database an attractive goal. Every three years, ICTV publishes a report listing viruses and their properties. Currently, 72 families of viruses have been classified and thousands of individual species have been identified and characterized to various extents. As pointed out by Maniloff, although the hundreds of pages of the ICTV report contain a great deal of data as well as ample cross references to other sources, the report does not incorporate data on individual species of virus.

TABLE 8.1   Computer Programs for Use in Sedimentation
Experiments, Scattering Studies, and Crystallography

Program	Ref.
Sedimentation Experiments	
ASYST	519
BIOSPIN	520
C++	517
CHARISMA	519
DLL	517
EQASSOC	521
EXCEL	519
FITPACK	522
Kaleidagraph	523
MLAB	521
MSTAR	525
NONLIN	520
ORIGIN	519
ORTHO	520
POLY	525
SVEDBERG	526
XLAEQ	524
XLAVEL (Beckman Instruments)	526
VELGAMMA (Beckman Instruments)	526
Scattering Studies	
Classical light scattering	
Aurora Wyatt Technologies (Santa Barbara, CA)	
Dynamic light scattering	271
X-ray and neutron scattering	412
Crystallography	
Virus crystallography	198
Crystals of macromolecules	527

"The ICTV report only defines genera and families. It's a consensus document that doesn't give data for individual viruses." By contrast, the universal database will include information "on real, observable virus particles." And it is being designed to be all-inclusive wherever possible. "The message was driven home to the subcommittee that, as long as we are going to the original literature, we should not omit data. For the most part, we must not be judges, but put in all available data and cite literature references."

In cases where inclusion of raw data would make the system prohibitively voluminous, plans call for providing accession numbers along with appropriate cross-references to other sources carrying the requisite data, such as databases containing DNA sequence information crystallogrpahic coordinates. (Source: *ASM News,* 60, 584–585, 1994.)

My intention is to develop a database on viruses to include mass-molecular weight values and related properties, such as sedimentation and diffusion coefficients, the partial specific volume, density, and viscosity.

8.3.2   Databases for Viral Components: Proteins and Nucleic Acids

Databases concerned with protein characterization include the Protein Database, a computer-based archival file for macromolecular structures, located at the Brookhaven National Laboratory; the PIR protein sequence database; and the Cambridge Crystallographic Database.[529]

For protein structural analyses, X-ray diffraction databases or derivatives of X-ray diffraction databases are necessary as a source of reference data for searching and matching the analyst's experimental data.

Diffraction databases fall into one of two categories: either a compilation of crystal interplanar spacings, d-spacings, giving information on the phase, formula, and crystal cell, and Miller indices, or a compilation of crystal cell data. The Powder Diffraction File (PDF) maintained by the International Centre for Diffraction Data (ICDD) is an example of a d-spacing-based database. The National Institute of Standards and Technology (NIST) Crystal Data File (CDF) is an example of a crystal cell database. The NIST/Sandia/ICDD Electron Diffraction Database is a derivative database, combining augmented information from the PDF and CDF, plus crystallographic and chemical information designed especially for application to electron diffraction search/match and related problems. Such databases generally include information on molecular weight values and related parameters.

Databases for nucleic acids are numerous and include GEnBank, located at the Los Alamos National Laboratory and Bethesda, Maryland. A relatively new facility is that of LIFESEQ, based in Palo Alto, California. LIFESEQ is a proprietary database holding tens of thousands of gene sequences extracted from tissue libraries.

In addition to well established computer systems allowing access to databanks such as PROPHET and Mosaic, LIMB (Listing of Molecular Biology databases) provides information on molecular biology and related databases.

Reprint Retrieval Systems — Often, the researcher needs a copy of the original study in order to access information. An outstanding document retrieval system is Knight-Ridder Source One (KR Source One) based in New York City. KR Source One has access to over 1.5 million periodical titles through its network of internal library collections and its external affiliates. In a matter of hours, the document/reprint can be located, photocopied (subject to royalty requirements), and made available to the requestor.

References

1. Bawden, F. C., *Plant Viruses and Virus Diseases*, Chronica Botanica Company, Waltham, MA, 1956.
2. Beijerinck, M. W., *A Contagium Vivum Fluidum as the Cause of the Mosaic Disease of Tobacco Leaves*, Verhandel, Akademie Wetenschap, Amsterdam, 6, 1, 1898.
3. Brock, T. D., *Milestones in Microbiology*, Prentice-Hall, Englewood Cliffs, NJ, 1961, 153.
4. Löffler, F. A. J. and Frosch, P., Report of the Committee for Research On the Foot-and-Mouth Disease, Part I, *Zentralb. Bakteriol., Parasit. Infekt.*, 23, 371–391, 1898.
5. Fenner, F., History of virology, in *Encyclopedia of Virology*, Vol. 2, Webster, R. G. and Granoff, A., Eds., Academic Press, 1994, 627–634.
6. Bergold, G. M., Insect viruses, *Adv. Virus Res.*, 1, 632, 1953.
7. Twort, F. W., An investigation on the nature of the ultramicroscopic viruses, *Lancet*, 11, 1241, 1915.
8. d'Herelle, F. H., An invisible microbe that is antagonistic to the dysentery bacillus, *C. R. Acad. Sci.*, 165, 373, 1917.
9. Knoll, M. and Ruska, E., The electron microscope, *Z. Phys.*, 78, 318, 1932.
10. Melnick, J. L., ECHO Viruses, in *Cellular Biology, Nucleic Acids, and Viruses*, Special Publications of the New York Academy of Sciences, Vol. 5, 1957, 367.
11. Lwoff, A. and Gutman, A., Production discontinue de bactériophages par une souche lysogène de bacillus megatherium, *C. R. Acad. Sci.*, 229, 679, 1949.
12. Cairns, J., Stent, G. S., and Watson, J. D., Eds., *Phage and the Origins of Molecular Biology*, expanded ed.,Cold Spring Harbor Press, Cold Spring Harbor, NY, 1992.
13. Zinder, N. D. and Lederberg, J., Genetic exchange in Salmonella, *J. Bacteriol.*, 64, 679, 1952.
14. Volkin, E., Astrachan, L., and Countryman, J. L., Metabolism of RNA phosphorus in E. coli infected with bacteriophage T7, *Virology*, 6, 545, 1958.
15. Brenner, S., Jacob, F., and Meselson, M., An unstable intermediate carrying information from genes to ribosomes for protein synthesis, *Nature*, 190, 575, 1961.
16. Hall, B. D. and Spiegelman, S., Sequence complementarity of T2 DNA and T2 specific RNA, *Proc. Natl. Acad. Sci. U.S.A.*, 47, 137, 1961.
17. Klug, W. S. and Cummings, M. R., *Concepts of Genetics*, 2nd ed., Merrill Publishing, Columbus, OH, 1986, 423–424.
18. Van Brunt, J., First hurdle in AIDS vaccine cleared, *Biotechnology*, 5, 1118, 1987.
19. Hershey, A. D. and Chase, M., Independent functions of viral protein and nucleic acid in growth of bacteriophage, *J. Gen. Physiol.*, 36, 39, 1952.
20. Gierer, A. and Schramm, G., Die infektiostat der ribonukleinsaure des tabakmosaikvirus, *Z. Naturforsch.*, 116, 138, 1956.
21. Fraenkel-Conrat, H., The role of the nucleic acid in the reconstitution of active tobacco mosaic virus, *J. Am. Chem. Soc.*, 78, 882, 1956.
22. Fraenkel-Conrat, H., Singer, B., and Williams, R. C., Infectivity of viral nucleic acid, *Biochem. Biophys. Acta*, 25, 87, 1957.

23. Avery, O. T., MacLeod, C. M., and McCarty, M., Studies on the chemical transformation of pneumococcal types, *J. Exp. Med.*, 79, 137, 1944.

24. Crick, F. C., On protein synthesis, *Symp. Soc. Exp. Biol.*, 12, 138, 1958.

25. Temin, H. M., The DNA provirus hypothesis: the establishment and implications of RNA-directed DNA synthesis, *Science*, 192, 1075, 1976.

26. Baltimore, D., Retroviruses and retrotransposons: the role of reverse transcription in shaping the eukaryotic genome, *Cell*, 40, 481, 1985.

27. Carey, J., O'C. Hamilton, J., Flynn, J., and Smith, G., The gene kings, *Business Week*, May 8, 1995, pp. 72–78.

28. Gibbs, A., Evolution of viruses, in *Encyclopedia of Virology*, Vol. 1, Webster, R. G. and Granoff, A., Eds., Academic Press, New York, 1994, 436–441.

29. Mills, D. R., Peterson, R. L., and Spiegelman, S., An extracellular Darwinian experiment with a self-duplicating nucleic acid molecule, *Proc. Natl. Acad. Sci. U.S.A.*, 58, 217, 1967.

30. Joyce, G. F., Directed molecular evolution, *Sci. Am.*, 267, 90, 1992.

31. Beaudry, A. A. and Joyce, G. F., Directed evolution of an RNA enzyme, *Science*, 257, 635, 1992.

32. Rous, P., Transmission of a malignant new growth by means of a cell-free filtrate, *JAMA*, 56, 198, 1911.

33. Rhoads, C. P., Viruses as causative agents in cancer, *Ann. NY Acad. Sci.*, 54, 859, 1952.

34. Pfister, H. and Fleckenstein, B., Tumor viruses — human, in *Encyclopedia of Virology*, Vol., 3, Webster, R. G. and Granoff, A., Eds., Academic Press, New York, 1994, 1492–1499.

35. Murphy, F. A., Fauquet, C. M., Bishop, D. H. L., Ghabrial, S. A., Jarvis, A. W., Martelli, G. P., Mayo, M. A., and Summers, M. D., *Virus Taxonomy, Sixth Report of the International Committee on Taxonomy of Viruses*, Springer-Verlag, New York, 1995.

36. Lister, R. M., Possible relationship of virus-specific products of tobacco rattle virus infections, *Virology*, 28, 350, 1966.

37. Lister, R. M., Functional relationships between virus-specific products of infection by viruses of the tobacco rattle type, *J. Gen. Virol.*, 2, 43, 1968.

38. Kassanis, B., Properties and behavior of a virus depending for its multiplication on another, *J. Gen. Microbiol.*, 27, 477, 1962.

39. Diener, T. O., Potato spindle tuber "virus". IV. A replicating low molecular weight RNA, *Virology*, 45, 411, 1971.

40. Prusiner, S. B., Novel proteinaceous infectious particles cause scrapie, *Science*, 216, 136, 1982.

41. Gajdusek, D. C., Unconventional viruses and the origin and disappearance of kuru, *Science*, 197, 943, 1977.

42. Muller, H. J., Physics in the attack on the fundamental problems of genetics, *Sci. Monthly*, 44, 210, 1936.

43. Bohr, N., Light and life, *Nature*, 131, 421, 1933.

44. Delbruck, M., A physicist looks at biology, *Conn. Acad. Sci.*, 38, 173, 1949.

45. Delbruck, M., quoted in Kay, L. E., *The Molecular Vision of Life. CalTech, The Rockefeller Foundation and the Rise of the New Biology*, Oxford University Press, New York, 1993, 135.

46. Judson, H. F., *The Eighth Day of Creation*, expanded ed., Cold Spring Harbor Press, Cold Spring Harbor, NY, 1996.

47. Timofeef-Ressovsky, N. W., Zimmer, K. G., and Delbruck, M., Natur der genmutation und der genstruktur, *Nachr. Ges. Wiss. Gottinge, Math-Phys. Kl. Fachgr.*, 6(1), 189, 1935.

48. Olby, R., *The Path to the Double Helix. The Discovery of DNA*, Dover Publications, New York, 1974, 232.

49. Schrödinger, E., *What is Life? The Physical Aspect of the Living Cell*, Cambridge University Press, New York, 1944.

50. Gamow, G., Possible reflection between deoxyribonucleic acid and protein structure, *Nature*, 173, 318, 1954.

51. Weaver, W., 'Molecular biology': origin of the term, *Science*, 170, 561, 1970, 'The Natural Sciences': Report of The Rockefeller Foundation, New York, 1938, 203–225.

52. Binnig, G., Quate, C. F., and Gerber, C., Atomic force microscope, *Europhys. Lett.,* 3, 1281, 1986.
53. Binnig, G., Rohrer, H., Gerber, C., and Weibel, E., Surface studies by scanning tunneling microscopy, *Phys. Rev. Lett.,* 49, 57,1982.
54. Haeberle, W., Hoerber, J. K. H., Ohnesorge, F., Smith, D. P., and Binnig, G., In situ investigation of single living cells infected by viruses, *Ultramicroscopy,* 42–44, 1161, 1992.
55. Lyubchenko, Y. L., Jacobs, B. L., and Lindsay, S. M., Atomic force microscopy of reovirus dsRNA: a routine technique for length measurements, *Nucleic Acids Res.,* 20, 3983, 1992.
56. Lyubchenko, Y. L., Oden, P. I., Lampner, D., Linday, S. M., and Dunker, K. A., Atomic force microscopy of DNA and bacteriophage in air, water, and propanol: the role of adhesion forces, *Nucleic Acids Res.,* 21, 1117, 1993.
57. Staudinger, H., Ueber die Konstitution des Kautschuks., *Ber. Chem. Ges.,* 57, 1203, 1924.
58. Weaver, W., Quoted in Kay, L. E., *The Molecular Vision of Life: Caltech, The Rockefeller Foundation, and the Rise of the New Biology,* Oxford University Press, New York, 1993, 134.
59. Luria, S. E., Delbruck, M., and Anderson, T. F., Electron microscope studies of bacterial viruses, *J. Bacteriol.,* 46, 57, 1943.
60. Kekulé, A., On the constitution and metamorphoses of chemical compounds and on the chemical nature of carbon, *Justus Liebig's Ann. Chem.,* 106, 129, 1858.
61. Kekulé, A., The scientific aims and achievements of chemistry, *Nature,* 18, 210, 1878.
62. Mark, H., *Giant Molecules,* Time, New York, 1966.
63. Brock, W. H., *The Norton History of Chemistry,* W.W. Norton, New York, 1992.
64. Conant, J. B. and Blatt, A. H., *The Chemistry of Organic Compounds,* 3rd ed., Macmillan, New York, 1948.
65. Fischer, E., Isomerie der polypeptide, *Sber. Preuss. Akad. Wiss. Halbd.,* 2, 990, 1916.
66. Werner, A., Ueber haupt-und nebenvalenzen und die constitution der ammonium verbindunge, *Justus Liebig's Ann.,* 322, 261, 1902.
67. Flory, P., *Principles of Polymer Chemistry,* Cornell University Press, Ithaca, NY, 1953.
68. Asimov, I., *Asimov's Biographical Encyclopedia of Science and Technology,* Doubleday, Garden City, NY, 1964.
69. Florkin, M., A history of biochemistry, *Compr. Biochem.,* 30, 279, 1972.
70. Sponsler, O. L. and Dore, W. H., The structure of ramie cellulose as derived from X-ray data, *Colloid Symp. Monogr.,* 4, 174, 1926.
71. Svedberg, T. and Rinde, H., The ultracentrifuge. A new instrument for the determination of size and distribution of size of particle in amicroscopic colloids, *J. Am. Chem. Soc.,* 46, 2677, 1924.
72. Lehninger, A. L., Nelson, D. L., and Cox, M. M., *Principles of Biochemistry,* 2nd ed., Worth Publishers, New York, 1993.
73. Loeb, J., *Proteins and the Theory of Colloidal Behavior,* McGraw-Hill, New York, 1922.
74. Troland, L. T., Biological enigmas and the theory of enzyme action, *American Naturalist,* 31, 321, 1917.
75. Ravin, A. W., The gene as a catalyst: the gene as an organism, *Stud. History Biol.,* 1, 1, 1977.
76. Woods, A. F., Inhibiting action of oxidase upon diastase, *Science,* 11, 17, 1900.
77. Hagedoorn, A. L., Autocatalytic substances: the determinants for the inheritable characters. A biochemical theory of inheritance and evolution, *Vortrage un Aufsatze Uber Entwickelungsmechanik der Organismen (Leipzig),* 12, 1, 1911.
78. Twort, F., An investigation on the nature of the ultramicroscopic viruses, *Lancet,* ii, 1241, 1915.
79. Northrop, J., Chemical nature and mode of formation of pepsin, trypsin, and bacteriophage, *Science,* 86, 479, 1937.
80. Mulvania, M., Studies on the nature of the virus of tobacco mosaic virus, *Phytopathology,* 16, 853, 1926.
81. Vinson, C. G. and Petri, A. W., Mosaic disease of tobacco. II. Activity of the virus precipitated by lead acetate. *Contrib. Boyce Thompson Inst.,* 3, 131, 1941.
82. Dvorak, M., The effect of mosaic on the globulin of potato, *J. Infect. Dis.,* 41, 215, 1927.

83. Purdy, H. A., Immunological reactions with tobacco mosaic virus, *J. Exp. Med.*, 49, 919, 1929.
84. Stanley, W. M., Isolation of a crystalline protein possessing the properties of tobacco mosaic virus, *Science*, 81, 644, 1935.
85. Kay, L. E., *The Molecular Vision of Life. Caltech, the Rockefeller Foundation, and the Rise of the New Biology*, Oxford University Press, New York, 1993, 111.
86. Stent, G. and Calendar, R., *Molecular Genetics*, W. H. Freeman, San Francisco, 1978, 578.
87. Bawden, F. C. and Pirie, N. W., The isolation and some properties of liquid, crystalline substances from solanaceous plants infected with three strains of tobacco mosaic virus, *Proc. R. Soc. London Ser. B.*, 123, 274, 1937.
88. Edsall, J. T., The development of the physical chemistry of proteins, 1898–1940, in *The Origins of Modern Biochemistry: A Retrospect on Proteins*, Srinivasan, P.R., Fruton, J. S., and Edsall, J.T., Eds., *Annals of the New York Academy of Sciences*, 325, 53, 1979.
89. Svedberg, T. and Rinde, H., The determination of the distribution of size of particles in disperse systems, *J. Am. Chem. Soc.*, 45, 943, 1923.
90. Svedberg, T. and Nichols, J. B., Determination of size and distribution of size of particle by centrifugal methods, *J. Am. Chem. Soc.*, 45, 2910, 1923.
91. Svedberg, T. and Fahraeus, R., A new method for the determination of the molecular weight of the proteins, *J. Am. Chem. Soc.*, 48, 430, 1926.
92. Adair, G. S., The osmotic pressure of hemoglobin in the absence of salts, *Proc. R. Soc. London Ser. A*, 108, 627, 1925.
93. Svedberg, T. and Chirnoaga, E., The molecular weight of hemocyanin, *J. Am. Chem. Soc.*, 50, 1399, 1928.
94. Pedersen, K. V., The Analytical Ultracentrifuge, The First Half Century, 1924–1974, Fractions (Beckman Instruments), No. 1, 1974.
95. Williams, J. W., The development of the ultracentrifuge and its contributions, in *The Origins of Modern Biochemistry: A Retrospect on Proteins*, Srinivasan, P. R., Fruton, J. S., and Edsall, J. T., Eds., *Annals of the New York Academy of Sciences*, 325, 77, 1979.
96. Branden, C. and Tooze, J., *Introduction to Protein Structure*, Garland Publishing, New York, 1991, 26.
97. Fruton, J. S. and Simmonds, S., *General Biochemistry*, John Wiley & Sons, New York, 1953.
98. Brow, H., Sanger, F., and Kitai, R., The structure of pig and sheep insulins, *Biochem. J.*, 60, 556, 1955.
99. Pauling, L. and Corey, R., Two hydrogen-bonded configurations of the polypeptide chain, *J. Am. Chem. Soc.*, 71, 5349, 1950.
100. Ramachandran, G. N. and Sassiekharan, V., Conformation of polypeptides and proteins, *Adv. Protein Chem.*, 28, 283, 1968.
101. Pauling, L., Corey, R. B., and Branson, H. R., The structure of proteins: two hydrogen-bonded configurations of the polypeptide chain, *Proc. Natl. Acad. Sci. U.S.A.*, 37, 205, 1951.
102. Fraenkel-Conrat, H., Protein chemists encounter viruses, in *The Origins of Modern Biochemistry, A Retrospect On Proteins*, Srinivasan, P. R., Fruton, J. S., and Edsall, J. T., Eds., *Annals of the New York Academy of Sciences*, 325, 309, 1979.
103. Jordan, D. O., *The Chemistry of Nucleic Acids*, Butterworths, Washington, D.C., 1960.
104. Davidson, J. N., *The Biochemistry of the Nucleic Acids*, 6th ed., Methuen, London, 1950.
105. Astbury, W. T. and Bell, F. O., Some recent developments in the X-ray study of proteins and related structures, *Cold Spring Harbor Symp. Quant. Biol.*, 6, 109, 1938.
106. Mudd, S., Contribution to the discussion, *Cold Spring Harbor Symp. Quant. Biol.*, 6, 118, 1938.
107. Furberg, S., Crystal structure of cytidine, *Nature*, 164, 22, 1949.
108. Franklin, R. E. and Gosling, R. G., Molecular structure of nucleic acids. Molecular configuration in sodium thymonucleate, *Nature*, 171, 740, 1953.
109. Wilkins, M. H. F., Stokes, A.R., and Wilson, H. R., Molecular structure of nucleic acids. molecular structure of deoxypentose nucleic acids, *Nature*, 171, 738, 1953.
110. Rich, A. A., Nordheim, A., and Wang, A. H. J., The chemistry and biology of left-handed Z-DNA, *Annu. Rev. Biochem.*, 53, 791, 1984.

111. Vischer, E. and Chargaff, E., The separation and quantitative estimation of purines and pyrimidines in minute amounts, *J. Biol. Chem.*, 176, 703, 1948.

112. Watson, J. D. and Crick, F. H. C., Molecular structure of nucleic acids. A structure for deoxyribose nucleic acid, *Nature*, 171, 737, 1953.

113. Astbury, W. T., Protein and virus studies in relation to the gene, *Int. Conf. Genet.*, 7, 49, 1939.

114. Eriksson-Quensel, I. B. and Svedberg, T., Sedimentation and electrophoresis of the tobacco mosaic virus protein, *J. Am. Chem. Soc.*, 58, 1863, 1936.

115. Lauffer, M. D., The molecular weight and shape of tobacco mosaic proteins, *Science*, 87, 469, 1938.

116. Takahashi, W. N. and Rawlins, T. E., Methods of determining shape of colloidal particles: application in study of tobacco mosaic virus, *Proc. Soc. Exp. Biol. Med.*, 30, 155, 1932.

117. Lauffer, M. D. and Stanley, W. M., Stream double refraction of virus proteins, *J. Biol. Chem.*, 123, 507, 1938.

118. Kausche, G. D., Pfankuch, E., and Ruska, H., Die Sichtbarmachung von Pflanzen virus im ubermikrosop, *Naturwissenschaften*, 18, 292, 1939.

119. Ruska, H., Ueber ein neues bei der bakteriophagen lyse im uebermikroskop, *Naturwissenschaften*, 29, 367, 1941.

120. Anderson, T. F., Electronmicroscopy of phages, in *Phage and the Origins of Molecular Biology*, expanded ed., Cairns, J., Stent, G. S., and Watson, J. D., Eds., Cold Spring Harbor Press, Cold Spring Harbor, New York, 1992, 63–78.

121. Luria, S. E. and Anderson, T. F., The identification and characterization of bacteriophages with the electron microscope, *Proc. Natl. Acad. Sci. U.S.A.*, 28, 127, 1942.

122. Anderson, T. F., Destruction of bacterial viruses by osmotic shock, *J. Appl. Phys.*, 21, 70, 1950.

123. Herriott, R. M., Nucleic acid-free T2 virus "ghosts" with specific biological action, *J. Bacteriol.*, 61, 252, 1951.

124. Williams, R. C. and Wyckoff, R. W. G., Electron shadow micrography of the tobacco mosaic virus protein, *Science*, 101, 594, 1945.

125. Brenner, S. and Horne, R. W., A negative staining method for high resolution electronmicroscopy of viruses, *Biochim. Biophys. Acta*, 34, 103, 1959.

126. Kassanis, B., Vince, D. A., and Woods, R. D., Light and electron microscopy of cells infected with tobacco necrosis and satellite viruses, *J. Gen. Virol.*, 7, 143, 1970.

127. Caspar, D. L. D. and Klug, A., Physical principles in the construction of regular viruses, *Cold Spring Harbor Symp. Quant. Biol.*, 27, 1, 1962.

128. Klug, A. and Caspar, D. L. D., The structure of small viruses, *Adv. Virus Res.*, 7, 225, 1960.

129. Fraenkel-Conrat, H. and Williams, R. C., Reconstitution of active tobacco mosaic virus from its inactive protein and nucleic acid components, *Proc. Natl. Acad. Sci. U.S.A.*, 41, 690, 1955.

130. Crick, F. H. C. and Watson, J. D., Structure of small viruses, *Nature (London)*, 177, 473, 1956.

131. Caspar, D. L. D., Structure of tomato bushy stunt virus, *Nature*, 177, 476, 1956.

132. Klug, A., Finch, J. T., and Franklin, R. E., The structure of turnip yellow mosaic virus: X-ray diffraction studies, *Biochim. Biophys. Acta*, 25, 242, 1957.

133. Williams, R. C. and Smith, K. M., The polyhedral form of the tipula iridescent virus, *Biochim. Biophys. Acta*, 28, 464, 1958.

134. Finch, J. T. and Klug, A., Structure of poliomyelitis virus, *Nature*, 183, 1709, 1959.

135. Klug, A. and Finch, J. T., The symmetries of the protein and nucleic acid in turnip yellow mosaic virus: X-ray diffraction studies, *J. Mol. Biol.*, 2, 201, 1960.

136. Caspar, D. L. D. and Holmes, K. C., Structure of the dahlemense strain of tobacco mosaic virus: a periodiocally deformed helix, *J. Mol. Biol.*, 46, 99, 1969.

137. Harrison, S., Olson, A. J., Schutt, C. E., Winkler, F. K., and Bricogne, G., Tomato bushy stunt virus at 2.9 Å., *Nature*, 276, 368, 1978.

138. Rossman, M. G., Arnold, E., Erickson, J. W., Frankenberger, E. A., Griffith, J. P., Johnson, J. E., Kramer, G., Luo, M., Mosser, A. C., Rueckert, R. R., Sherry, B., and Vriend, G., Structure of a human cold virus and functional relationship to other picornaviruses, *Nature*, 317, 145, 1985.

139. Hogle, J., Chow, M., and Filman, D. J., Three-dimensional structure of polio viruses at 2.9 Å resolution, *Science,* 229, 1358, 1985.

140. Acharya, R., Fry, E., Stuart, D. I., Fox, G., Rowlands, D., and Brown, F., The three-dimensional structure of foot-and-mouth disease virus at 2.9 Å., *Nature,* 337, 709, 1989.

141. Valegard, K., The three dimensional structure of the bacterial virus MS2, *Nature,* 344, 36, 1990.

142. Stanley, W. M. and Wyckoff, R. W. G., The isolation of tobacco ringspot and other virus proteins by ultracentrifugation, *Science,* 85, 181, 1937.

143. Brakke, M. K., Density gradient centrifugation. A new separation technique, *J. Am. Chem. Soc.,* 73, 1847, 1951.

144. Harvey, E. N., The tension at the surface of marine eggs, especially those of the sea urchin, *Biol. Bull.,* 61, 273, 1931.

145. Meselson, M., Stahl, F. W., and Vinograd, J., Equilibrium sedimentation of macromolecules in density gradients, *Proc. Natl. Acad. Sci. U.S.A.,* 43, 581, 1957.

146. Bock, R. M. and Ling, N.S., Devices for gradient elution in chromatography, *Anal. Chem.,* 26, 1543, 1954.

147. Anderson, N. G., The development of zonal centrifuges and ancillary systems for tissue fractionation and analysis, *Natl. Cancer Inst. Monogr.,* 21, 1, 1966.

148. Seaborg, G. T., *Natl. Cancer Inst. Monogr.,* 21, ix, 1966.

149. Anderson, N. G., An introduction to particle separations in zonal centrifuges, *Natl. Cancer Inst. Monogr.,* 21, 9, 1966.

150. Anderson, N. G., Harris, W. W., Barber, A. A., Rankin, C. T., Jr., and Candler, E. G., Separation of subcellular components and viruses by combined rate and isopycnic zonal centrifugation, *Natl. Cancer Inst. Monogr.,* 21, 253, 1966.

151. Anderson, N. G., Waters, D. A., Nunley, C. E., Gibson, R. F., Schilling, R. M., Denny, E. C., Cline, G. B., Babelay, E. F., and Perardi, T. E., K-Series centrifuges. I. Development of the K-II continuous-sample-flow-with-banding centrifuge system for vaccine purification, *Anal. Biochem.,* 32, 460, 1966.

152. Wright, R. R., Pappas, W. S., Carter, J. A., and Weber, C. W., Preparation and recovery of compounds for density gradient solutions, *Natl. Cancer Inst. Monogr.,* 21, 241, 1966.

153. Mazzone, H. M., Breillatt, J. P., and Anderson, N. G., Zonal rotor purification and properties of a nuclear polyhedrosis virus of the European pine sawfly (*Neodiprion sertifer,* Geoffroy), in *Proc. IV Int. Colloq. Insect Pathology,* Bickley, W. E., Ed., University of Maryland Press, College Park, MD, 1970, 371.

154. Giddings, J. C., A new separation concept based on a coupling of concentration and flow nonuniformities, *Separation Sci.,* 1, 123, 1966.

155. Arner, E. C. and Kirkland, J. J., Sedimentation field flow fractionation, in *Analytical Ultracentrifugation in Biochemistry and Polymer Science,* Harding, S. E., Rowe, A. J., and Horton, J. C., Eds., Royal Society of Chemistry, Cambridge, U.K., 1992, chap. 12.

156. Fitch, R. M., *Polymer Colloids. A Comprehensive Introduction,* Academic Press, San Diego, 1997, 110–113.

157. Giddings, J. C., Caldwell, K. D., and Kesner, L. F., Molecular weight distribution from field-flow fractionation, in *Determination of Molecular Weight,* Cooper, A. R., Ed., John Wiley & Sons, New York, 1989, chap. 12.

158. Giddings, J. C., Field-flow fractionation: analysis of macromolecular, colloidal, and particulate materials, *Science,* 260, 1456, 1993.

159. Caldwell, K. D., Nguyen, T. T., Giddings, J. C., and Mazzone, H. M., Field flow fractionation of alkali-liberated nuclear polyhedrosis from gypsy moth, *Lymantria dispar* Linnaeus, *J. Virol. Methods,* 1, 241, 1980.

160. Yonker, C. R., Caldwell, K. D., Giddings, J. C., and Van Etten, J. L., Physical characterization of PBCV virus by sedimentation field flow fractionation, *J. Virol. Methods,* 11, 145, 1985.

161. Hünefeld, F. L., *Die Chemismus in der Thierischen Organisation,* R.A. Brockhaus, Leipzig, 1840, 160.

162. Schultz, F. N., *Die Krystallisation von Eiweisstoffen und ihre Bedeutung für die Eiweisschemie,* Gustav Fischer, Jena, 1901.

163. Osborne, T. B., *The Vegetable Proteins*, 2nd ed., Longmans Green, London, 1924.

164. Sumner, J. B., The isolation and crystallization of the enzyme urease, *J. Biol. Chem.*, 69, 435, 1926.

165. Northrop, J. H. and Kunitz, M., Crystalline trypsin. I. Isolation and tests of purity, *J. Gen. Physiol.*, 16, 267, 1932.

166. Bawden, F. C., Pirie, N. W., Bernal, J. D., and Fankuchen, I., Liquid crystalline substances from virus-infected plants, *Nature (London)*, 138, 1051, 1936.

167. Bernal, J. D. and Fankuchen, I., Structure types of protein "crystals" from virus-infected plants, *Nature (London)*, 139, 923, 1937.

168. Friedrich, W., Knipping, P., and v. Laue, M., Interferenz-erscheinungen bei rongenstrahlen, *Sitz. Ber. Math. Phys. Klasse, Bayer Akad. Wiss.*, Munich, 1912, 303.

169. Bragg, W. H., X-rays and crystal structure, *Philos. Trans. R. Soc. A.*, 215, 253, 1915.

170. Asimov, I., *Asimov's Biographical Encyclopedia of Science and Technology*, Doubleday, Garden City, NY, 1964.

171. Bragg, W. L., The determination of parameters in crystal structures by means of Fourier series, *Proc. R. Soc. London Ser. A*, 123, 537, 1929.

172. Hodgkin, D. C., Crytallographic measurements and the structure of protein molecules, in *The Origins of Modern Biochemistry: A Retrospect on Proteins*, Srinivasan, P. R., Fruton, J. S., and Edsall, J. T., Eds., Annals of the New York Academy of Science, Vol. 25, 1979, 123–125.

173. Hutchison, A., Memorandum on the history, development and present condition of the Department of Mineralogy, *Cambridge University Reporter*, Dec. 10, 390, 1929.

174. Astbury, W. T. and Street, A., X-ray studies of the structure of hair, wool and related fibers, *Philos. Trans. R. Soc. A*, 230, 75, 1931.

175. Astbury, W. T. and Woods, H. J., X-ray studies of the structure of hair, wool and related fibers. II. The molecular structure and elastic properties of hair keratin, *Philos. Trans. R. Soc. A*, 232, 333, 1934.

176. Bernal, J. D. and Crowfoot, D., X-ray photographs of crystalline pepsin, *Nature*, 133, 794, 1934.

177. Olby, R., *The Path to the Double Helix. The Discovery of DNA*, Dover Publications, New York, 1974, 254.

178. Huggins, M. L., *Physical Chemistry of High Polymers*, John Wiley & Sons, New York, 1958.

179. Crowfoot, D. G. and Schmidt, G. M. J., X-ray crystallographic measurements on a single crystal of a tobacco necrosis virus derivative, *Nature*, 155, 505, 1945.

180. Bernal, J. D. and Carlisle, C. H., Unit cell measurements of wet and dry crystalline turnip yellow mosaic virus, *Nature*, 162, 139, 1948.

181. Carlisle, C. H. and Dornberger, K., Some X-ray measurements on single crystals of tomato bushy stunt virus, *Acta Crystallogr.*, 1, 194, 1948.

182. Perutz, M., Hemoglobin structure and respiratory transport, *Sci. Am.*, 239 (6), 92, 1978.

183. Kendrew, J. C., The three dimensional structure of a protein molecule, *Sci. Am.*, 205, 96, 1961.

184. Franklin, R. E. and Gosling, R. G., The structure of sodium thymonucleate fibres, I. The influence of water content, *Acta Crystallogr.*, 6, 673, 1953.

185. Wilkins, M. H. F. and Randall, J. T., Crystallinity in sperm heads, molecular structure of nucleoprotein in vivo, *Biochim. Biophys. Acta*, 10, 192, 1953.

186. Branden, C. and Tooze, J. *Introduction to Protein Structure*, Garland Publishing, New York, 1991, 269–270.

187. Bernal, J. D., The crystal structure of the natural amino acids and related compounds, *Z. Krystallogr. Miner.*, 78, 363, 1931.

188. Branden, C. and Tooze, J., *Introduction to Protein Structure*, Garland Publishing, New York, 1991, 270.

189. Borchardt-Ott, W., *Crystallography*, 2nd ed., Springer, Berlin, 1995.

190. Day, J. and McPherson, A., Macromolecular crystal growth experiments on International Microgravity Laboratory-I, *Protein Sci.*, 1, 1254, 1992.

191. McPherson, A., Virus and protein crystal growth on earth and in microgravity, *J. Phys. D Appl. Phys.*, 26, B104, 1993.

192. Valverde, R. A. and Dodds, J. A., Some properties of isometric virus particles which contain the satellite RNA of tobacco mosaic virus, *J. Gen. Virol.,* 68, 965, 1987.

193. Patterson, A. L., A direct method for the determination of the components of interatomic distances in crystals, *Z. Kristallogr.,* 90, 517, 1935,

194. Branden, C. and Tooze, J., *Introduction to Protein Structure,* Garland Publishing, New York, 1991, 276.

195. Rhodes, G., *Crystallography Made Crystal Clear, A Guide for Macromolecular Models,* Academic Press, New York, 1993, 27.

196. Silva, A. M. and Rossman, M. G., The refinement of southern bean mosaic virus in reciprocal space, *Acta Crytallogr.,* B, 41, 147, 1985.

197. Branden, C. and Tooze, J., *Introduction to Protein Structure,* Garland Publishing, New York, 1991, 277.

198. Fry, E., Logan, D., and Stuart, D., Virus crystallography, in *Crystallographic Methods and Protocols.* Jones, C., Mulloy, B., and Sanderson, M. R., Eds., *Methods in Molecular Biology,* Vol. 56, Humana Press, Totowa, NJ, 1996, chap. 13.

199. Rossman, M. G. and Blow, D. M., The detection of subunits within the crystallographic asymmetric unit, *Acta Crystallogr.,* 15, 24–31, 1962.

200. Bricogne, G., Methods and programs for direct-space exploitation of geometric redundancies, *Acta Crystallogr. A,* 32, 832–847, 1976.

201. Harrison, S. C., Olson, A. J., Schutt, C. E., Winkler, F. K., and Bricogne, G., Tomato bushy stunt virus at 2.9 Å resolution, *Nature,* 276, 368–373, 1978.

202. Abad-Zapatero, C., Abdel-Meguid, S., Johnson, J. E., Leslie, A. G. W., Rayment, I., Rossman, M.G., Suck, D., and Tsukihara, T., Structure of southern bean mosaic virus at 2.8Å resolution, *Nature,* 286, 33–39, 1980.

203. Lijas, L., Unge, T., Jones, T. A., Fridborg, K., Lovgren, S., Skoglund, U., and Strandberg, B., Structure of satellite tobacco necrosis virus at 3.0 Å resolution, *J. Mol. Biol.,* 159, 93–108, 1982.

204. Arnold, E., Vriend, G., Luo, M., Griffith, J. P., Kramer, G., Erickson, J. W., Johnson, J. E., and Rossman, M. G., The structure determination of a common cold virus, human rhinovirus 14, *Acta Crystallogr. A,* 43, 346, 1987.

205. Hogle, J. M. Chow, M., and Filman, D. J., Three-dimensional structure of poliovirus at 2.9 Å resolution, *Science,* 229, 1358–1365, 1985.

206. Valegard, K., Liljas, L., Fridborg, K., and Unge, T., The three-dimensional structure of the bacterial virus MS2, *Nature,* 344, 36–41, 1990.

207. McKenna, R., Xia, D., Willingham, P., Ilag, L. L., Krishnaswamy, S., Rossman, M. G., Olson, N. H., Baker, T. S., and Incardona, N. L., Structure of single-stranded DNA bacteriophage φX 174 and its functional implications, *Nature,* 355, 137–143, 1992.

208. Tsao, J., Chapman, M. S., Agbandje, M., Keller, W., Smith, K., Wu, H., Luo, M., Smith, T. J., Rossman, M. G., Compans, R. W., and Parrish, C. R., The three-dimensional structure of canine parvovirus and its functional implications, *Science,* 251, 1456–1464, 1991.

209. Liddington, R. C., Yan, Y., Moulai, J., Sahli, R., Benjamin, T. L., and Harrison, S. C., Structure of Simian virus 40 at 3.8 Å resolution, *Nature,* 354, 278–284, 1992.

210. Mizuno, H., Omura, T., Koisumi, M., Kondoh, M., and Tsukihara, T., Crystallization and preliminary X-ray study of a double-shelled spherical virus, rice dwarf virus, *J. Mol. Biol.,* 219, 665–669, 1991.

211. Basak, A. K., Grimes, J., Burroughs, J. N., Mertons, P. P. C., Roy, P., and Stuart, D., Crystallographic studies on the structure of the bluetongue virus inner capsid, in *Bluetongue, African Horse Sickness and Related Orbiviruses,* Walton, T. E. and Osburn, B.L., Eds., CRC Press, Boca Raton, FL, 1992, 483–490.

212. Basak, A. K., Stuart, D. I., and Roy, P., Preliminary crystallographic study of bluetongue virus capsid protein VP7, *J. Mol. Biol.,* 228, 687–689, 1992.

213. Gong, Z. X., Wu, H., Cheng, R. H., Hull, R., and Rossman, M. G., Crystallization of cauliflower mosaic virus, *Virology,* 179, 941–945, 1990.

214. Coombs, K. M., Fields, B. N., and Harrison, S. C., Crystallization of the reovirus type 3 Dearing core. Crystal packing is determined by the λ2 protein, *J. Mol. Biol.*, 215, 1–5, 1990.

215. Chen, Z., Stuffacher, C., Li, Y., Schmidt, T., Bomu, W., Kamer, G., Shanks, M., Lomonossoff, G., and Johnson, J. E., Protein-RNA interactions in an icosahedral virus at 3.0 Å resolution, *Science*, 245, 154–159, 1989.

216. Hosur, M. V., Schmidt, T., Tucker, R. C., Johnson, J. E., Gallagher, T. M., Selling, B. H., and Rueckert, R. R., Structure of an insect virus at 3.0 Å resolution, *Proteins*, 2 (3), 167, 1987.

217. Acharya, R., Fry, E., Stuart, D., Fox, G., Rowlands, D., and Brown, F., The three-dimensional structure of foot-and-mouth disease virus at 2.9 Å resolution, *Nature*, 337, 709, 1989.

218. Luo, M., Vriend, G., Kamer, G., Minor, I., Arnold, E., Rossman, M. G., Boege, U., Scraba, D. G., Duke, G. M., and Palmenberg, A. C., The atomic structure of mengo virus at 3.0 Å resolution, *Science*, 235, 182, 1987.

219. Larson, S. B., Koszelak, S., Day, J., Greenwood, A., Dodds, J. A., and McPherson, A., Three-dimensional structure of satellite tobacco mosaic virus at 2.9 Å resolution, *J. Mol. Biol.*, 231, 375, 1993.

220. Jones, T. A. and Liljas, L., Structure of satellite tobacco necrosis virus after crystallographic refinement at 2.5 Å resolution, *J. Mol. Biol.*, 177, 735, 1984.

221. Subramanya, H. S., Gopinath, K., Nayudu, M. V., Savithri, H. S., and Murthy, M. R., Structure of sesbania mosaic virus at 4.7 Å resolution and partial sequence of the coat protein, *J. Mol. Biol.*, 229, 20, 1993.

222. Hogle, J. M., Maeda, A., and Harrison, S. C., Structure and assembly of turnip crinkle virus. I. X-ray crystallographic structure analysis at 3.2 Å resolution, *J. Mol. Biol.*, 191, 625, 1986.

223. Lechner, M.D., Determination of molecular weight averages and molecular weight distributions from sedimentation equilibrium, in *Analytical Ultracentrifugation in Biochemistry and Polymer Science*, Harding, S. E., Rowe, A. J., and Horton, J. C., Eds., Royal Society of Chemistry, Cambridge, U.K., 1992, chap. 16.

224. Svedberg, T. and Pedersen, K. O., *The Ultracentrifuge*, Clarendon Press, Oxford, 1940.

225. Giebler, R., The Optima XL-A: A new analytical ultracentrifuge with a novel precision absorption optical system, in *Analytical Ultracentrifugation in Biochemistry and Polymer Science*, Harding, S. E., Rowe, A. J., and Horton, J. C., Eds., Royal Society of Chemistry, Cambridge, U.K., 1992, 16.

226. Svedberg, T. and Rinde, H., The determination of the distribution of size of particles in disperse systems, *J. Am. Chem. Soc.*, 45, 943, 1923.

227. Ralston, G., *Introduction to Ultracentrifugation*, Beckman Instruments, Palo Alto, CA, 1993.

228. Vinograd, J., Bruner, R., Kent, R., and Weigle, J., Band centrifugation of macromolecules and viruses in self-generating density gradients, *Proc. Natl. Acad. Sci. U.S.A.*, 49, 902, 1963.

229. Belli, M., The evaluation of standard sedimentation coefficient in band sedimentation of macromolecules, *Biopolymers*, 12, 1853, 1973.

230. Stafford, W. F., Jansco, A., and Graceffa, P., Caldesmon from rabbit liver: molecular weight and length by analytical ultracentrifugation, *Arch. Biochem. Biophys.*, 281, 66, 1990.

231. Lebowitz, J., Kar, S., Braswell, E., McPherson, S., and Richard, D. L., Human immunodeficiency virus-1 reverse transcriptase heterodimer stability, *Protein Sci.*, 3, 1374, 1994.

232. Lebowitz, J., Stability of the Human Immunodeficiency Virus-1 Reverse Transcriptase Heterodimer, Application Information, Beckman Instruments, Palo Alto, CA, 1995.

233. Schachman, H. K. and Edelstein, S. J., Ultracentrifugal studies with absorption optics and a split-beam photoelectric scanner, *Methods Enzymol.*, 27, 3, 1973.

234. Chervenka, C. H., *A Manual of Methods for the Analytical Ultracentrifuge*, Beckman Instruments, Palo Alto, CA, 1969.

235. Trautman, R., Operating and comparing procedures facilitating schlieren pattern analysis in analytical ultracentrifugation, *J. Phys. Chem.*, 60, 1211, 1956.

236. LaBar, F. E. and Baldwin, R. L., A study by interference optics of sedimentation in short columns, *J. Phys. Chem.*, 66, 152, 1962.

237. Rowe, A. J., Wynne-Jones, S., Thomas, D. G., and Harding, S. E., Methods for off-line analysis of sedimentation velocity and sedimentation equilibrium patterns, in *Analytical Ultracentrifugation in Biochemistry and Polymer Science*, Harding, S. E., Rowe, A. J., and Horton, J. C., Eds., Royal Society of Chemistry, Cambridge, U.K., 1992, chap. 5.

238. Yphantis, D. A., Lary, J. W., Stafford, W. F., Liu, S., Olsen, P. H., Hayes, D. B., Moody, T., Ridgeway, T. M., Lyons, D. A., and Laue, T. M., On line data acquisition for the rayleigh interference optical system of the analytical ultracentrifuge, in *Modern Analytical Ultracentrifugation*, Schuster, T. M. and Laue, T. M., Eds., Birkhauser, Boston, 1994, 209.

239. De Rosier, D. J., Munk, P., and Cox, D. J., Automatic measurement of interference photographs for the ultracentrifuge, *Anal. Biochem.*, 50, 139, 1971.

240. Schachman, H. K., Is there a future for the ultracentrifuge?, in *Analytical Ultracentrifugation in Biochemistry and Polymer Science*, Harding, S. E., Rowe, A. J., and Horton, J. C., Eds., Royal Society of Chemistry, Cambridge, U.K., 1992, chap. 1.

241. Gray, R. A., Stern, A., Bewley, T., and Shire, S. J., Rapid Determination of Spectrophotometric Absorptivity by Analytical Ultracentrifugation. Application Information, Beckman Instruments, Palo Alto, CA, 1995.

242. Schumaker, V. and Rees, A., Preparative centrifugation in virus research, in *Principles and Techniques in Plant Virology*, Kado, C. I. and Agrawal, H. O., Eds., Van Nostrand Reinhold, New York, 1972, chap. 12.

243. Trautman, R. and Hamilton, M. G., Analytical ultracentrifugation, in *Principles and Techniques in Plant Virology*, Kado, C. I. and Agrawal, H. O., Eds., Van Nostrand Reinhold, 1972, chap. 18.

244. Furst, A., Overview of Sedimentation Velocity for the Optima XL-A Analytical Ultracentrifuge, Technical Information DS-819, Beckman Instruments, Palo Alto, CA, 1991.

245. Schachman, H. K., *Ultracentrifugation in Biochemistry*, Academic Press, New York, 1959.

246. Trautman, R. and Schumaker, V., Generalization of the radial dilution square law in ultracentrifugation, *J. Chem. Phys.*, 22, 551, 1954.

247. Rowe, A. J., The concentration dependence of sedimentation, in *Analytical Ultracentrifugation in Biochemistry and Polymer Science*, Royal Society of Chemistry, Cambridge, U.K., 1992, chap. 21.

248. Johnston, J. P. and Ogston, A. G., A boundary anomaly found in the ultracentrifugal sedimentation of mixtures, *Trans. Faraday Soc.*, 42, 789, 1946.

249. Gilbert, G. A. and Gilbert, L. M., Determination in the ultracentrifuge of protein heterogeneity by computer modeling, illustrated by pyruvate dehydrogenase multienzyme complex, *J. Mol. Biol.*, 144, 405, 1980.

250. Stafford, W. F., Boundary analysis in sedimentation transport experiments: a procedure for obtaining sedimentation coefficient distributions using the time derivative of the concentration profile, *Anal. Biochem.*, 203, 295, 1992.

251. Stafford, W. F., Sedimentation boundary analysis of interacting systems: use of the apparent sedimentation coefficient distribution, in *Modern Analytical Ultracentrifugation*, Schuster, T. M. and Laue, T. M., Eds., Birkhauser, Boston, 1994, 119.

252. Stafford, W. F., Methods of boundary analysis in sedimentation velocity experiments, in *Numerical Computer Methods*, Part B, *Methods in Enzymology*, Vol. 240, Johnson, M. L. and Brand, L., Eds., Academic Press, New York, 1994.

253. Stafford, W. F., Methods of obtaining sedimentation coefficient distributions, in *Analytical Ultracentrifugation in Biochemistry and Polymer Science*, Harding, S. E., Rowe, A. J., and Horton, J. C., Eds., Royal Society of Chemistry, Cambridge, U.K., 1992, 359.

254. Liu, S. and Stafford, W. F., A real-time video-based Rayleigh optical system for an analytical ultracentrifuge allowing imaging of the entire centrifuge cell, *Biophys. J.*, 61, A476, 2745, 1992.

255. Stafford, W. F., Sedimentation boundary analysis: an averaging method for increasing the precision of the Rayleigh optical system by nearly two orders of magnitude, *Biophys. J.*, 61, A476, #2746, 1992.

256. Dawes, E. A., *Quantitative Problems in Biochemistry*, E.& S. Livingstone, Edinburgh, 1956.

257. Mazzone, H. M., Biophysical and Biochemical Studies of Squash Mosaic Virus and Related Macromolecules, Ph.D. thesis, University of Wisconsin, Madison, 1959.

258. Comper, W. D. and Preston, B. N., The analytical ultracentrifuge as a tool for diffusion measurements. Cross diffusion effects in ternary polymer: polymer: solvent systems, in *Analytical Ultracentrifugation in Biochemistry and Polymer Science*, Harding, S. E., Rowe, A. J., and Horton, J. C., Eds., Royal Society of Chemistry, Cambridge, U.K., 1992, chap. 23.

259. Creeth, J. M. and Pain, R. H., The determination of molecular weights of biological macromolecules by ultracentrifuge methods, *Prog. Biophys. Mol. Biol.*, 17, 217, 1967.

260. Schumaker, V. N. and Schachman, H. K., Ultracentrifugal analysis of dilute solutions, *Biochim. Biophys. Acta*, 23, 628, 1957.

261. Bloomfield, V. A. and Lim, T. K., Quasi-elastic laser light scattering, *Methods Enzymol.*, 28, 415, 1978.

262. Harding, S. E., Determination of diffusion coefficients of biological macromolecules by dynamic light scattering, in *Methods in Molecular Biology*, Vol. 22, *Microscopy, Optical Spectroscopy and Macroscopic Techniques*, Jones, C., Mulloy, B., and Thomas, A. H., Eds., Humana Press, Totowa, NJ, 1994, 97.

263. Pecora, R., Doppler shifts in light scattering from pure liquids and polymer solutions, *J. Chem. Phys.*, 40, 1604, 1964.

264. Cummins, H. Z., Knabel, N., and Yeh, Y., Observation of diffusion broadening of rayleigh scattered light, *Phys. Rev. Lett.*, 12, 150, 1964.

265. Dubin, S. B., Lunacek, J. H., and Benedek, G. B., Observation of the spectrum of light scattered by solutions of biological macromolecules, *Proc. Natl. Acad. Sci. U.S.A.*, 57, 1164, 1967.

266. Dubin, S. B., Benedek, G. B., Bancroft, F. C., and Freifelder, D., Molecular weights of coliphages and coliphage DNA. II. Measurement of diffusion coefficients using optical mixing spectroscopy, and measurement of sedimentation coefficients, *J. Mol. Biol.*, 54, 547, 1970.

267. Markham, R., Diffusion, in *Methods in Virology*, Vol. 2, Maramorosch, K. and Koprowski, H., Eds., Academic Press, New York, 1967, 275.

268. Burchard, W., Static and dynamic light scattering approaches to structure determination of biopolymers, in *Laser Light Scattering in Biochemistry*, Harding, S. E., Sattelle, D. B., and Bloomfield, V. A., Eds., Royal Society of Chemistry, Cambridge, U.K., 1992, 3.

269. Godfrey, R. E., Johnson, P., and Stanley, C. J., The intensity fluctuation spectroscopy method and its application to viruses and larger enzymes, in *Biomedical Applications of Laser Light Scattering*, Sattelle, D. B., Lee, W. I., and Ware, B. R., Eds., Elsevier, Amsterdam, 1982, 373.

270. Brisco, M., Haniff, C., Hull, R., Wilson, T. M. A., and Sattelle, D. B., The kinetics of swelling of southern bean mosaic virus: a study using photon correlation spectroscopy, *Virology*, 148, 218, 1986.

271. Johnsen, R. M. and Brown, W., An overview of current methods of analyzing QLS data, in *Laser Light Scattering in Biochemistry*, Harding, S. E., Sattelle, D. B., and Bloomfield, V. A., Eds., Royal Society of Chemistry, Cambridge, U.K., 1992, 77.

272. Pusey, P. N. and Vaughn, J. M., Light scattering and intensity fluctuation spectroscopy, in *Dielectric and Related Molecular Processes*, Vol. 2, Davies, M., Ed., The Chemical Society, London, 1975, 48.

273. Provencher, S., Low-bias macroscopic analysis of polydispersity, in *Laser Light Scattering in Biochemistry*, Harding, S. E., Sattelle, D. B., and Bloomfield, V. A., Eds., Royal Society of Chemistry, Cambridge, U.K., 1992, 92.

274. Johnson, P. and Brown, W., An investigation of rigid rod-like particles, in *Laser Light Scattering in Biochemistry*, Harding, S. E., Sattelle, D. B., and Bloomfield, V. A., Eds., Royal Society of Chemistry, Cambridge, U.K., 1992, 161.

275. Rarity, J. G., Owens, P. C. M., Atkinson, T., Seabrook, R. N., and Carr, R. J. G., Light scattering studies of protein association, in *Laser Light Scattering in Biochemistry*, Harding, S. E., Sattelle, D. B., and Bloomfield, V. A., Eds., Royal Society of Chemistry, Cambridge, U.K., 1992, 240.

276. Koppel, D. E., Study of *Escherichia coli* by intensity fluctuation spectroscopy of scattered laser light, *Biochemistry*, 13, 2712, 1974.

277. Lowenstein, M. A. and Birnboim, M. H., A method for measuring sedimentation and diffusion of macromolecules in capillary tubes by total intensity and quasi-elastic light-scattering techniques, *Biopolymers,* 14, 419, 1975.

278. Bellamy, A. R., Gillies, S.C., and Harvey, J. D., Molecular weight of two Oncornavirus genomes. Derivation from particle molecular weights and RNA content, *J. Virol.,* 14, 1388, 1974.

279. Ware, B. R., Raj, T., Flygare, W. H., Lesnaw, J. A., and Reichman, M. E., Molecular weights of vesicular stomatitis virus and its defective particles by laser light beating spectroscopy, *J. Virol.,* 11, 141, 1973.

280. Schaefer, D. W., Benedek, G. B., Schofield, P., and Bradford, E., Spectrum of light quasielastically scattered from tobacco mosaic virus, *J. Chem. Phys.,* 55, 3884, 1971.

281. Dubos, P., Hallett, R., Kells, D. T. C., Sorensen, O., and Rowe, D., Biophysical studies of infectious pancreatic necrosis virus, *J. Virol.,* 22, 150, 1977.

282. Camerini-Otero, R. D., Pusey, P. N., Koppel, D. E., Schaefer, D. W., and Franklin, R. M., Intensity fluctuation spectroscopy of laser light scattered by solutions of spherical viruses: R17, QB, BSV, PM2, and T7. II. Diffusion coefficients, molecular weights, solvation, and particle dimensions, *Biochemistry,* 13, 960, 1974.

283. Fujita, H., *Foundations of Ultracentrifugal Analysis,* John Wiley & Sons, New York, 1975.

284. Fujita, H., Notes on the derivation of sedimentation equilibrium equations, in *Modern Analytical Ultracentrifugation,* Schuster, T. M. and Laue, T. M., Eds., Birkhauser, Boston, 1994, 3.

285. Goldberg, R. J., Sedimentation in the ultracentrifuge, *J. Phys. Chem.,* 57, 194, 1953.

286. Wills, P. R. and Winzor, D. J., Thermodynamic non-ideality and sedimentation equilibrium, in *Analytical Ultracentrifugation in Biochemistry and Polymer Science,* Harding, S. E., Rowe, A. J., and Horton, J. C., Eds., Royal Society of Chemistry, Cambridge, U.K., 1992, 311.

287. Lansing, W. D. and Kraemer, E. O., Molecular weight analysis of mixtures by sedimentation equilibrium in the Svedberg ultracentrifuge, *J. Am. Chem. Soc.,* 57, 1369, 1935.

288. Fujita, H., *Mathematical Theory of Sedimentation Analysis,* Academic Press, New York, 1962.

289. Condino, J., The Determination of Molecular Weights by Sedimentation Equilibrium, Technical Information, Beckman Instruments, Palo Alto, CA, 1992.

290. Harding, S. E., Determination of absolute molecular weights using sedimentation equilibrium analytical ultracentrifugation, in *Methods in Molecular Biology,* Vol. 22, *Microscopy, Optical Spectroscopy, and Macroscopic Techniques,* Jones, C., Mulloy, B., and Thomas, A. H., Eds., Human Press, Totowa, NJ, 1994, 75.

291. Rowe, A. J. and Harding, S. E., Off-line schlieren and Rayleigh data capture for sedimentation velocity and equilibrium analysis, in *Analytical Ultracentrifugation in Biochemistry and Polymer Science,* Harding, S. E., Rowe, A. J., and Horton, J. C., Eds., Royal Society of Chemistry, Cambridge, U.K., 1992, 49.

292. Laue, T., On-line data acquisition and analysis from the Rayleigh interferometer, in *Analytical Ultracentrifugation in Biochemistry and Polymer Science,* Harding, S. E., Rowe, A. J., and Horton, J. C., Eds., Royal Society of Chemistry, Cambridge, U.K., 1992, 63.

293. Harding, S. E. and Johnson, P., Physicochemical studies on turnip yellow mosaic virus: homogeneity, molecular weights, hydrodynamic radii and concentration dependence parameters, *Biochem. J.,* 231, 549, 1985.

294. Richards, E. G. and Schachman, H. K., Ultracentrifuge studies with Rayleigh interference optics. I. General applications, *J. Phys. Chem.,* 63, 1578, 1959.

295. Yphantis, D. A., Equilibrium ultracentrifugation of dilute solutions. *Biochemistry,* 3, 297, 1964.

296. Correia, J. J. and Yphantis, D. A., Equilibrium sedimentation in short solution columns, in *Analytical Ultracentrifugation in Biochemistry and Polymer Science,* Harding, S. E., Rowe, A. J., and Horton, J. C., Eds., Royal Society of Chemistry, Cambridge, U.K., 1992, chap. 13.

297. Mason, M. and Weaver, W., The settling of small particles in a fluid, *Phys. Rev.,* 23, 412, 1924.

298. Van Holde, K. E. and Baldwin, R. L., Rapid attainment of sedimentation equilibrium, *J. Phys. Chem.,* 62, 734, 1958.

299. Yphantis, D. A., Rapid determination of molecular weights of peptides and proteins, *Ann. NY Acad. Sci.*, 88, 586, 1960.

300. Beauchamp, K. G., *Walsh Functions and Their Applications*, Academic Press, New York, 1975.

301. Schmid, K. and Mazzone, H. M., Determination of the particle weight of spherical viruses, *Nature*, 197, 671, 1963.

302. Archibald, W. J., A demonstration of some new methods of determining molecular weights from the data of the ultracentrifuge, *J. Phys. Colloid Chem.*, 51, 1204, 1947.

303. Bellamy, A. R., Gillies, S. C., and Harvey, J. D., Molecular weight of two Oncornavirus genomes. Derivation from particle molecular weights and RNA content, *J. Virol.*, 14, 1388, 1974.

304. Fuscaldo, A. A., Aaselstad, H. G., and Hoffman, E. J., Biological, physical, and chemical properties of eastern equine encephalitis virus. I. Purification and physical properties, *J. Virol.*, 7, 233, 1971.

305. Burness, A. T. H. and Clothier, F. W., Particle weight and other biophysical properties of encephalomyocarditis virus, *J. Gen. Virol.*, 6, 381, 1970.

306. Salmeen, I., Rimai, L., Luftig, R. B., Liebes, L., Retzel, E., Rich, M., and McCormick, J. J., Hydrodynamic diameters of murine mammary, Rous sarcoma, and feline leukemia RNA tumor viruses: studies by laser beat frequency light-scattering spectroscopy and electron microscopy, *J. Virol.*, 17, 584, 1976.

307. Vande Woude, G. P., Swaney, J. B., and Bachrach, H. L., Chemical and physical properties of foot-and-mouth disease virus: a comparison with Maus Elberfeld virus, *Biochem. Biophys. Res. Commun.*, 48, 1222, 1972.

308. Dobos, P., Hill, B. J., Hallett, R., Kells, D. T. C., Becht, H., and Teninges, D., Biophysical and biochemical characterization of five animal viruses with bisegmented double-stranded RNA genomes, *J. of Virol.*, 32, 593, 1979.

309. Hoyle, L., *The Influenza Viruses*, Virology Monogr. 4, 1968, Springer, New York.

310. Salzman, L. A. and Jori, L. A., Characteristics of Kilham rat virus, *J. Virol.*, 5, 114, 1970.

311. Marvin, D. A. and Hoffmann-Berling, H., A fibrous DNA phage (fd) and a spherical RNA phage (fr) specific for male strains of *E. coli.*, *Z. Naturforsch.*, 18b, 884, 1963.

312. Schwerdt, C. E., Physical and chemical characteristics of purified poliomyelitis virus, in *Cellular Biology, Nucleic Acids and Viruses*, Special Publication of the New York Academy of Sciences, 5, 157, 1957.

313. Frisque, R. J., Barbanti-Brodano, G., Crawford, L. V., Gardner, S. D., Howley, P. M., Ortho, G., Shah, K. V., Vader Noordaa, J., and Zur Hausen, H., Papovaviridae, in *Virus Taxonomy*, Sixth Report of the International Committee on Taxonomy of Viruses, Murphy, F. A., Fauquet, C. M., Bishop, D. H. L., Ghabrial, S. A., Jarvis, A. W., Martelli, G. P., Mayo, M. A., and Summers, M. A., Eds., Springer-Verlag, New York, 1995, 136.

314. Salmeen, I., Rimai, L., Liebes, L., Rich, M. A., and McCormick, J. J., Hydrodynamic diameters of RNA tumor viruses. Studies by laser beat frequency of light scattering spectroscopy of avian myeloblastosis and Rauscher murine leukemia viruses, *Biochemistry*, 14, 134, 1975.

315. Gomatos, P. J. and Tamm, I., The secondary structure of reovirus RNA, *Proc. Natl. Acad. Sci. U.S.A.*, 49, 707, 1963.

316. McGregor, S. and Mayor, H. D., Biophysical and biochemical studies on rhinovirus and poliovirus. II. Chemical and hydrodynamic analysis of the rhinovirion, *J. Virol.*, 7, 41, 1971.

317. Medappa, K. C., McLean, C., and Rueckert, R. R., On the structure of rhinovirus 1A, *Virology*, 44, 259, 1971.

318. Fraenkel-Conrat, H., *The Chemistry and Biology of Viruses*, Academic Press, 1969, New York.

319. Boettger, M. and Scherneck, S., Heterogeneity, molecular weight and stability of an oncogenic papovavirus of the Syrian hamster, *Arch. Geschwulst. Forsch.*, 55, 225, 1985.

320. Ware, B. R., Raj, T., Flygare, W. H., Lesnaw, J. A., and Reichmann, M. E., Molecular weights of vesicular stomatitis virus and its defective particles by laser light scattering spectroscopy, *J. Virol.*, 11, 142, 1973.

321. Kaplan, R. L., Yelton, D. B., and Gerencser, V. F., Biochemical and biophysical properties of hyphomicrobium bacteriophage Hy φ30, *J. Virol.*, 19, 899, 1976.

322. Marvin, D. A. and Hohn, B., Filamentous bacterial viruses, *Bacteriol. Rev.*, 33, 172, 1969.

323. Dyson, R. D. and Van Holde, K. E., An investigation of bacteriophage lambda, its protein ghosts and subunits, *Virology*, 33, 559, 1967.

324. Möller, W. J., Determination of diffusion coefficents and molecular weights of ribonucleic acids and viruses, *Proc. Natl. Acad. Sci. U.S.A.*, 51, 501, 1964.

325. Schito, G. C., Rialdi, G., and Pesce, A., Biophysical properties of N4 coliphage, *Biochim. Biophys. Acta*, 129, 482, 1966.

326. Day, L. A. and Mindich, L., The molecular weight of bacteriophage φ6 and its nucleocapsid, *Virology*, 103, 376, 1980.

327. Sinsheimer, R. L., Purification and properties of bacteriophage φX174, *J. Mol. Biol.*, 1, 37, 1959.

328. Brown, D.T., Brown, N. C., and Burlingham, B. T., Morphology and physical properties of staphylococcus bacteriophage P11–M15, *J. Virol.*, 9, 664, 1972.

329. Velikodvorskaya, G. A., Bolkova, A. F., Petrovsky, G.V., and Tikhonenko, T. I., Physical and chemical properties of phage PB-2, *Vop. Virusol.*, 17, 332, 1972.

330. Pitout, M. J., Conradie, J. D., and Van Rensburg, A. J., Relationship between the sedimentation coefficient and molecular weight in bacteriophages, *J. Gen. Virol.*, 4, 577, 1969.

331. Taylor, N. W., Epstein, H. T., and Lauffer, M. A., Particle weight, hydration, and shape of T2 bacteriophage of *Escherichia coli*, *J. Am. Chem. Soc.*, 77, 1270, 1955.

332. Swanby, L. S., Some biophysical properties of the T3 bacteriophage of *Escherichia coli*, *Diss. Abstr.* 20, 1567, 1959.

333. Goldwasser, E. and Putnam, F., Physicochemical properties of T6 bacteriophage, *J. Biol. Chem.*, 190, 61, 1952.

334. Saunders, G. F. and Campbell, L. L., Characterization of thermophilic bacteriophage for Bacillus stearothermophilis, *J. Bacteriol.*, 91, 340, 1966.

335. Van Etten, J. L., Lane, L. C., and Meints, R. H., Unicellular plants also have large dsDNA viruses, *Sem. Virol.*, 2, 71, 1991.

336. Kelly, D.C., Barwise, A. H., and Walker, I. O., DNA contained by two densonucleosis viruses, *J. Virol.*, 21, 396, 1977.

337. Bancroft, J. B., Purification and properties of bean pod mottle virus and associated centrifugal and electrophoretic components, *Virology*, 16, 419, 1962.

338. Yamazaki, H., Bancroft, J., and Kaesberg, P., Biophysical studies of broad bean mottle virus, *Proc. Natl. Acad. Sci. U.S.A.*, 47, 979, 1961.

339. Bockstahler, L. E. and Kaesberg, P., The molecular weight and other biophysical properties of bromegrass mosaic virus, *Biophys. J.*, 2, 1, 1962.

340. Tremaine, J. H., Physical, chemical, and serological studies on carnation mottle virus, *Virology*, 42, 611, 1970.

341. Bancroft, J. B., Hiebert, E., Rees, M.W., and Markham, R., Properties of cowpea chlorotic mottle virus: its protein and nucleic acid, *Virology*, 34, 224, 1968.

342. Tremaine, J. H., Purification and properties of cucumber necrosis virus and a smaller top component, *Virology*, 48, 582, 1972.

343. Mazzone, H. M., Incardona, N. L., and Kaesberg, P., Biochemical and biophysical properties of squash mosaic virus and related macromolecules, *Biochim. Biophys. Acta*, 55, 164, 1962.

344. Markham, R., The biochemistry of plant virus, in *The Viruses*, Vol. 2, Burnet, F. M. and Stanley, W. M., Eds., Academic Press, New York, 1959, 33.

345. Hersh, R. T. and Schachman, H. K., On the size of the protein subunits in bushy stunt virus, *Virology*, 6, 234, 1958.

346. Kassanis, B., Properties and behaviour of a virus depending for its multiplication on another, *J. Gen. Microbiol.*, 27, 477, 1962.

347. Harrison, B. D. and Klug, A., Relationship between length and sedimentation coefficient for particles of tobacco rattle virus, *Virology*, 30, 738, 1966.

348. Steere, R. L., Purification and properties of tobacco ringspot virus, *Phytopathology*, 46, 60, 1956.

349. Haselkorn, R., Studies on infectious RNA from turnip yellow mosaic virus, *J. Mol. Biol.*, 4, 357, 1962.

350. Howlett, G. J., The preparative ultracentrifuge as an analytical tool, in *Analytical Ultracentrifugation in Biochemistry and Polymer Science*, Harding, S. E., Rowe, A. J., and Horton, J. C., Eds., Royal Society of Chemistry, Cambridge, U.K., 1992, 32.

351. Howlett, G. J., Sedimentation analysis of membrane proteins, in analytical ultracentrifugation in biochemistry and polymer science, Harding, S. E., Rowe, A. J., and Horton, J. C., Eds., Royal Society of Chemistry, U.K., 1992, 470.

352. Pickels, E. G., Sedimentation in the angle centrifuge, *J. Gen. Physiol.*, 26, 341, 1943.

353. Friedewald, W. F. and Pickels, E. G., Centrifugation and ultrafiltration studies in allantoic fluid preparations of influenza virus, *J. Exp. Med.*, 79, 301, 1944.

354. Kahler, H. and Lloyd, B. J., Jr., Density of polystyrene latex by a centrifugal method, *Science*, 114, 34, 1951.

355. Hogeboom, G. H. and Kuff, E. L., Sedimentation behavior of proteins and other materials in a horizontal preparative rotor, *J. Biol. Chem.*, 210, 733, 1954.

356. deDuve, C. and Berthet, J., The use of differential centrifugation in the study of tissue enzymes, *Int. Rev. Cytol.*, 3, 225, 1954.

357. Brakke, M., Density gradient centrifugation. a new separation technique, *J. Am. Chem. Soc.*, 73, 1847, 1951.

358. Freifelder, D., Zonal centrifugation, *Methods Enzymol.*, 27 (Part D), 140, 1973.

359. Schumaker, V. N. and Rosenbloom, J., Fundamental mass-transport equations for zone sedimentation velocity, *Biochemistry*, 4, 1005, 1965.

360. Brakke, M. K., Estimation of sedimentation constants of viruses by density-gradient centrifugation, *Virology*, 6, 96, 1958.

361. Anderson, N. G., Studies on isolated cell components. VIII. High resolution centrifugation, *Exp. Cell Res.*, 9, 446, 1955.

362. Linderstrom-Lang, K. and Lanz, H., Jr., Studies on enzymatic histochemistry. XXIX. Dilatometric microestimation of peptidase activity, *C. R. Trav. Lab. Carlsberg (Chim.)*, 21, 315, 1938.

363. Meselson, M. and Stahl, F. W., The replication of DNA in E. coli, *Proc. Natl. Acad. Sci. U.S.A.*, 44, 671, 1958.

364. Meselson, M., Stahl, F. W., and Vinograd, J., Equilibrium centrifugation of macromolecules in density gradients, *Proc. Natl. Acad. Sci. U.S.A.*, 4, 581, 1957.

365. Baldwin, R. L., Equilibrium sedimentation in a density gradient of materials having a continuous distribution of effective densities, *Proc. Natl. Acad. Sci. U.S.A.*, 45, 939, 1959.

366. Baldwin, R. L. and Van Holde, K. E., Sedimentation of high polymers, *Fortschr. Hochpolym-Forsch, Bd.*, 1, 451, 1960.

367. Ift, J. B. and Vinograd, J., The buoyant behavior of bovine serum mercaptalbumin in salt solutions at equilibrium in the ultracentrifuge. II. Net hydration, ion binding, and solvated molecular weight in various salt solutions, *J. Phys. Chem.*, 70, 2814, 1966.

368. Chervenka, C.H. and Elrod, L. H., A *Manual of Methods for Large Scale Centrifugation*, Beckman Instruments, Palo Alto, CA, 1972.

369. Bishop, B. S., Digital computation of sedimentation coefficients in zonal centrifuges, in *The Development of Zonal Centrifuges and Ancillary Systems for Tissue Fractionation and Analysis*, Anderson, N. G., Ed., *Natl. Cancer Inst. Monogr.*, 21, 175, 1966.

370. Vinograd, J., Bruner, R., Kent, R., and Weigle, J., Band-centrifugation of macromolecules and viruses in self-generating density gradients, *Proc. Natl. Acad. Sci. U.S.A.*, 49, 902, 1963.

371. Stafford, W. F., Jansco, A., and Graceffa, P., Caldesmon from rabbit liver: molecular weight and length by analytical ultracentrifugation, *Arch. Biochem. Biophys.*, 281, 66, 1990.

372. Liebermann, H. and Mentel, R., Quantification of adenovirus particles, *J. Virol. Methods*, 50, 281, 1994.

373. de Verdugo, U. R., Selinka, H-C., Huber, M., Kramer, B., Kellermann, J., Hofschneider, P. H., and Kandolf, R., Characterization of a 100-kilodalton binding protein for the six serotypes of coxsackie B viruses, *J. Virol.*, 69, 6751, 1995.

374. Grasser, F. A., Haiss, P., Gottel, S., and Mueller-Lantzsch, N., Biochemical characterization of Epstein-Barr virus nuclear antigen 2A, *J. Virol.*, 65, 3779, 1991.

375. Schneemann, A., Zhong, W., Gallagher, T. M., and Rueckert, R. R., Maturation cleavage required for infectivity of a nodavirus, *J. Virol.*, 66, 6728, 1992.

376. Schneemann, A., Dasgupta, R., Johnson, J. E., and Rueckert, R. R., Use of recombinant baculoviruses in synthesis of morphologically distinct viruslike particles of flock house virus, a nodavirus, *J. Virol.*, 67, 2756, 1993.

377. Ohlinger, V. F., Haas, B., Myers, G., Weiland, F., and Thiel, H.-J., Identification and characterization of the virus causing rabbit hemorrhagic disease, *J. Virol.*, 64, 3331, 1990.

378. Curry, S., Chow, M. and Hogle, J.M., The poliovirus 135S particle is infectious, *J. Virol.*, 70, 7125, 1996.

379. Reynolds, C., Birnby, D., and Chow, M., Folding and processing of the capsid protein precursor P1 is kinetically retarded in neutralization site 3B mutants of poliovirus, *J. Virol.*, 66, 1641, 1992.

380. Shepard, D. A., Heinz, B. A., and Rueckert, R. R., WIN 52035-2 inhibits both attachment and eclipse of human rhinovirus 14, *J. Virol.*, 67, 2245, 1993.

381. Lee, W-M., Monroe, S. S., and Rueckert, R. R., Role of maturation clevage in infectivity of picornaviruses: activation of an infectosome, *J. Virol.*, 67, 2110, 1993.

382. Laue, T. M., Shah, B. D., Ridgeway, T. M., and Pelletier, S. L., Computer-aided interpretation of analytical sedimentation data for proteins, in *Analytical Ultracentrifugation in Biochemistry and Polymer Science*, Harding, S. E., Rowe, A. J., and Horton, J. C., Eds., Royal Society of Chemistry, Cambridge, U.K., 1992, 90.

383. Casassa, E. T. and Eisenberg, H., Thermodynamic analysis of multicomponent systems, *Adv. Protein Chem.*, 19, 287, 1964.

384. Durchschlag, H., Specific volumes of biological macromolecules and some other molecules of biological interest, in *Thermodynamic Data for Biochemistry and Biotechnology*, Hinz, H.-J., Ed., Springer-Verlag, New York, 1986, chap. 3.

385. Schachman, H. K., Ultracentrifugation, diffusion, and viscometry, *Methods Enzymol.*, 4, 32, 1957.

386. Bauer, N., Determination of density, *Techniques Org. Chem.*, 1, 253, 1949.

387. Vinograd, J. and Hearst, J. E., Equilibrium sedimentation of macromolecules and viruses in a density gradient, *Fortschr. Chem. Org. Naturstoff.*, 20, 372,1962.

388. Cohn, E. J. and Edsall, J. T., *Proteins, Amino Acids and Peptides as Ions and Dipolar Ions*, Reinhold, New York, 1943, 370.

389. Perkins, S. J., Protein volumes and hydration effects. The calculation of partial specific volumes, neutron scattering matchpoints and 280-nm absorption coefficients for proteins and glycoproteins from amino acid sequences, *Eur. J. Biochem.*, 157, 169, 1986.

390. Freifelder, D., Molecular weights of coliphages and coliphage DNA. IV. Molecular weights of DNA from bacteriophages T4, T5, and T7 and general problems of determination of M, *J. Mol. Biol.*, 54, 567, 1970.

391. Bancroft, F. C. and Freifelder, D., Molecular weights of coliphages and coliphage DNA. I. Measurement of the molecular weight of bacteriophage T7 by high-speed equilibrium centrifugation, *J. Mol. Biol.*, 54, 537, 1970.

392. Svedberg, T. and Eriksson-Quensel, I.-B., Haemocyanin in heavy water, *Nature*, 137, 400, 1936.

393. Edelstein, S. J. and Schachman, H. K., The simultaneous determination of partial specific volumes and molecular weights with microgram quantities, *J. Biol. Chem.*, 242, 306, 1967.

394. Daniels, F. and Alberty, R. D., *Physical Chemistry*, John Wiley & Sons, New York, 1955, 503–506.

395. Harding, S. E., Determination of macromolecular homogeneity, shape, and interactions using sedimentation velocity analytical ultracentrifugation, in *Microscopy, Optical Spectroscopy, and Macroscopic Techniques*, Jones, C., Mulloy, B., and Thomas, A. H., Eds., Humana Press, Totowa, NJ, 1994, chap. 5.

396. Lauffer, M. A., The molecular weight and shape of tobacco mosaic virus protein, *Science*, 87, 469, 1938.

397. Lauffer, M. A., The viscosity of tobacco mosaic virus protein solutions, *J. Biol. Chem.*, 126, 443, 1938.

398. Ball, D. A., Solution viscosity of polymers: an overview, *Am. Lab.*, 29, 24, 1997.

399. Schito, G. C., Rialdi, G., and Pesce, A., Biophysical properties of N4 coliphage., *Biochim. Biophys. Acta*, 129, 482, 1966.

400. Pitout, M. J., Conradie, J. D., and Van Rensburg, A. J., Relationship between the sedimentation coefficient and molecular weight of bacteriophages, *J. Gen. Virol.*, 4, 577, 1969.

401. Gerencser, V. F. and Voelz, H., A bacteriophage active on *Hyphomicrobium*, *Virology*, 44, 631, 1971.

402. Arner, E. C. and Kirkland, J. J., Sedimentation field flow fractionation, in *Analytical Ultracentrifugation in Biochemistry and Polymer Science*, Harding, S.E., Rowe, A.J., and Horton, J.C., Royal Society of Chemistry, Cambridge, U.K., 1992, 208.

403. Caldwell, K. D., Polymer analysis by field-flow fractionation, in *Modern Methods of Polymer Characterization*, Barth, H. G. and Mays, J. W., Eds., John Wiley & Sons, New York, 1991, chap. 4.

404. Caldwell, K. D., Karaiskakis, G., and Giddings, J. C., Characterization of T4D virus by sedimentation field flow fractionation, *J. Chromatogr.*, 215, 323, 1981.

405. Sklaviadis, T., Dreyer, R., and Manuelidis, L., Analysis of Creutzfeldt-Jakob disease infectious fractions by gel permeation chromatography and sedimentation field flow fractionation, *Virus Res.*, 26, 241, 1992.

406. Kirkland, J. J., Yau, W. W., Doerner, W. A., and Grant, J. W., Sedimentation field flow fractionation of macromolecules and colloids, *Anal. Chem.*, 52, 1944, 1980.

407. Caldwell, K. D., Nguyen, T. T., Giddings, J. C., and Mazzone, H. M., Field flow fractionation of alkali-liberated nuclear polyhedrosis virus from gypsy moth, *Lymantria dispar* Linnaeus, *J. Virol. Methods*, 1, 241, 1980.

408. Yonker, C. R., Caldwell, K. D., Giddings, J. C., and Van Etten, J. L., Physical characterization of PBCV virus by sedimentation field flow fractionation, *J. Virol. Methods*, 11, 145, 1984.

409. Harding, S. E., Berth, G., Ball, A., Mitchell, J. R., and de la Torre, J. G., The molecular weight distribution and conformation of citrus pectins in solution studied by hydrodynamics, *Carbohydrate Polymers*, 16, 1, 1991.

410. Harding, S. E., Sedimentation analysis of polysaccharides, in *Analytical Ultracentrifugation in Biochemistry and Polymer Science*, Harding, S. E., Rowe, A. J., and Horton, J. C. Eds., Royal Society of Chemistry, Cambridge, U.K., 1992, 495.

411. Schmidt, B. and Riesner, D., A fluorescent detection system for the analytical ultracentrifuge and its application to proteins, nucleic acids, viroids, and viruses, in *Analytical Ultracentrifugation in Biochemistry and Polymer Science*, Harding, S. E., Rowe, A. J., and Horton, J. C., Eds., Royal Society of Chemistry, Cambridge, U.K., 1992, 176.

412. Perkins, S. J., High-flux X-ray and neutron solution scattering, in *Microscopy, Optical Spectroscopy, and Macroscopic Techniques*, Vol. 22, Jones, C., Mulloy, B., and Thomas, A. H., Eds., Humana Press, Totowa, NJ, 1994, 39

413. Kratky, O., Low-angle scattering in polymers, *J. Polym. Sci.*, 3, 195, 1948.

414. Leonard, B. R., Jr., Anderegg, J. W., Kaesberg, P., Shulman, S., and Beeman, W. W., On the size, shape, and hydration of Southern bean mosaic virus and tobacco necrosis virus in solution, *J. Chem. Phys.*, 19, 793, 1951.

415. Pessen, H., Kumosinski, T. F., and Timasheff, S. N., Small-angle X-ray scattering, in *Methods Enzymol.*, 27, 209, 1973.

416. Debye, P., Molecular weight determination by light scattering, *J. Phys. Colloid Chem.*, 51, 18, 1947.

417. Guinier, A., La diffraction des rayons X aux tre petit angles: application a l'etude de phenomenes ultramicroscopique, *Ann. Phys.*, 12, 161, 1939.

418. Pollard, E. G., *The Physics of Viruses*, Academic Press, New York, 1953.

419. Nave, C., Helliwell, J. R., Moore, P. R., Thompson, A. W., Worgan, J. S., Greenall, R. J., Miller, A., Burley, S. K., Bradshaw, J., Pigram, W. J., Fuller, W., Siddons, D. P., Deutsch, M., and Tregear, R. T., Facilities for solution scattering and fibre diffraction at the Daresbury SRS, *J. Appl. Crystallogr.*, 18, 396, 1985.

420. Towns-Andrews, E., Berry, A., Bordas, J., Mant, G. R., Murray, P. K., Roberts, K., Sumner, I., Worgan, J. S., Lewis, R., and Gabriel, A., A time-resolved X-ray diffraction station: X-ray optics, detectors and data acquisition, *Rev. Sci. Instrum.*, 60, 2346, 1989.

421. Perkins, S. J., X-ray and neutron scattering, *New Compr. Biochem.*, 18B (Part II), 143, 1988.

422. Zipper, P., Kratky, O., Herrman, R., and Hohn, T., An X-ray small angle study of the bacteriophages fr and R17, *Eur. J. Biochem.*, 18, 1, 1971.

423. Jacrot, B. and Zaccai, G., Determination of molecular weight by neutron scattering, *Biopolymers*, 20, 2413, 1981.

424. Richards, R. W., Neutron and X-ray scattering, in *Determination of Molecular Weight*, Cooper, A. R., Ed., John Wiley & Sons, New York, 1989, chap. 6.

425. Guinier, A. and Fournet, G., *Small Angle Scattering of X-rays*, John Wiley & Sons, New York, 1955.

426. Hammouda, B., Krueger, S., and Glinka, C. J., Small angle neutron scattering at the National Institute of Standards and Technology, *J. Res. NIST*, 98, 31, 1993.

427. Timmins, P. A. and Zaccai, G., Low resolution structures of biological complexes studied by neutron scattering, *Eur. Biophys. J.*, 15, 257, 1988.

428. Jacrot, B., Chauvin, C., and Witz, J., Comparative neutron small-angle scattering study of small spherical RNA viruses, *Nature*, 266, 417, 1977.

429. Chauvin, C., Witz, J., and Jacrot, B., Structure of the tomato bushy stunt virus: a model for protein-RNA interaction, *Biologia*, 124, 641, 1978.

430. Kruse, J.,Timmins, P. A., and Witz, J., A neutron scattering study of the structure of compact and swollen forms of southern bean mosaic virus, *Virology*, 119, 42, 1982.

431. Devaux, C., Timmins, P. A., and Berthet-Colominas, C., Structural studies of adenovirus type 2 by neutron and X-ray scattering, *J. Mol. Biol.*, 167, 119, 1983.

432. Cusak, S., Neutron scattering studies of virus structure, in *Neutrons in Biology*, Schoenborn, B. P., Ed., Plenum Press, New York, 1984, 173.

433. Chauvin, C., Jacrot, B., and Witz, J., The structure and molecular weight of satellite tobacco necrosis virus: a neutron small-angle scattering study, *Virology*, 83, 479, 1977.

434. Freeman, R. and Leonard, K. R., Comparative mass measurement of biological macromolecules by scanning transmission electron microscopy, *J. Microscopy*, 122, 275, 1980.

435. Cuillel, M., Tripier, F., Braunwald, J., and Jacrot, B., A low resolution structure of frog virus 3, *Virology*, 99, 277, 1979.

436. Torbet, J., Neutron scattering study of the solution structure of bacteriophages Pf1 and fd, *FEBS Lett.*, 108, 61, 1979.

437. Harding, S. E., Classical light scattering for the determination of absolute molecular weights and gross conformation of biological macromolecules, in *Microscopy, Optical Spectroscopy, and Macroscopic Techniques*, Jones, C., Mulloy, B., and Thomas, A. H., Eds., chap. 7, Human Press, Totowa, NJ, 1994.

438. Pittz, E. P., Lee, J. C., Bablouzian, B., Townend, R., and Timasheff, S. N., Light scattering and differential refractometry, *Methods Enzymol.*, 27 (Part D), 209, 1973.

439. Zimm, B. H., Apparatus and methods for measurement and interpretation of angular variation of light scattering: preliminary results on polystyrene solutions, *J. Chem. Phys.*, 16, 1099, 1948.

440. Johnson, C. S., Jr. and Gabriel, D. A., *Laser Light Scattering*, Dover Publications, New York, 1981.

441. Yang, J. T., An improvement in the graphic treatment of angular light scattering data, *J. Polym. Sci.*, 26, 305, 1957.

442. Harding, S. E., Applications of light scattering in microbiology, *Biotech. Appl. Biochem.*, 8, 489, 1986.

443. Martinsen, A., Skjåk-Braek, G., Smidsrød, O., Zanetti, F., and Paoletti, S., Comparison of different methods for determination of molecular weight and molecular weight distribution of alginates, *Carbohydr. Polym.*, 15, 171, 1991.

444. Corona, A. and Rollings, J. E., Polysaccharide characterization by aqueous size exclusion chromatography and low angle light scattering, *Separation Size Technol.*, 23, 855, 1988.

445. Johnson, P. and McKenzie, G. H., A laser light scattering study of haemoglobin systems, *Proc. R. Soc. London Ser. B*, 199, 263, 1977.

446. Jackson, C., Nilsson, L. M., and Wyatt, P. J., Characterization of biopolymers using a multi-angle light scattering detector with size-exclusion chromatography, *J. Appl. Polym. Sci. Appl. Polym. Symp.*, 43, 99, 1989.

447. Wyatt, P. J., Combined differential light scattering with various liquid chromatography separation techniques, *Laser Light Scattering in Biochemistry*, Harding, S. E., Sattelle, D. B., and Bloomfield, V. A., Eds., Royal Society of Chemistry, Cambridge, U.K., 1992, 35.

448. Stacey, K. A., *Light Scattering in Physical Chemistry*, Academic Press, New York, 1956, 86.

449. Strauss, J. H. and Sinsheimer, R. L., Purification and properties of bacteriophage MS2 and of its ribonucleic acid, *J. Mol. Biol.*, 7, 43, 1963.

450. Overby, L. R., Barlow, G. H., Doi, R. H., Jacob, M., and Spiegelman, S., Comparison of two serologically distinct ribonucleic acid bacteriophages. I. properties of the viral particles, *J. Bacteriol.*, 91, 442, 1966.

451. Podzimek, S., The use of GPC coupled with a multiangle laser light scattering photometer for the characterization of polymers. On the determination of molecular weight, size, and branching, *J. Appl. Polym. Sci.*, 54, 91, 1994.

452. Jeng, L., Balke, S. T., Mourey, T. H., Wheeler, L., and Romeo, P., Evaluation of light scattering detectors for size exclusion chromatography. I. Instrument precision and accuracy, *J. Appl. Polym. Sci.*, 48, 1359, 1993.

453. Kasparkova, V. and Ommundsen, E., Determination of molar mass and radius of gyration by size exclusion chromatography with on-line viscometer and multi-angle laser light scattering. High temperature characterization of polystyrene, *Polymer*, 34, 1765, 1993.

454. Milas, M., Rinaudo, M., and Borsali, R., Radius of gyration and intrinsic viscosity of polyelectrolyte solutions. Role of the ionic strength, *Cien. Cult.*, 45, 46, 1993.

455. Reed, W. F., Data evaluation for unified multi-detector size exclusion chromatography — molar mass, viscosity and radius of gyration distributions, *Macromol. Chem Phys.*, 196, 1539, 1995.

456. Wyatt, P. J. and Shortt, D. W., Light Scattering Viscometry, International GPC Symposium '91, Waters Division of Millipore Filter Company, Bedford, MA, 1991.

457. Roessner, D. and Kulicke, W., On-line coupling of flow field-flow fractionation and multi-angle laser light scattering, *J. Chromatogr. A*, 687, 249, 1994.

458. Wen, J., Arakawa, T., and Philo, J. S., Size-exclusion chromatography with on-line light-scattering, absorbance, and refractive index detectors for studying proteins and their interactions, *Anal. Biochem.*, 240, 155, 1996.

459. Haney, M. A., Gillespie, D., and Yau, W. W., Viewing polymer structure through the triple "Lens" of SEC, *Today's Chemist at Work*, December 1994, 39–43.

460. Harding, S. E., Sattelle, D. B., and Bloomfield, V. A., Eds., *Laser Light Scattering in Biochemistry*, Royal Society of Chemistry, Cambridge, U.K., 1991.

461. Zeitler, E. and Bahr, G. F., A photometric procedure for weight determination of submicroscopic particles, *Quant. Electron Microscopy, J. Appl. Phys.*, 33, 847, 1962.

462. Bahr, G. F. and Zeitler, E., The determination of dry mass in populations of isolated particles, *Lab. Invest.*, 14, 142, 1965.

463. Lampert, F., Bahr, G. F., and Rabson, A. S., Herpes simplex virus: dry mass, *Science*, 166, 1163, 1969.

464. Hall, C. E., *Introduction to Electron Microscopy*, McGraw-Hill, New York, 1953, chap. 2.

465. Bahr, G. F., Determination of the dry mass of small biological objects by quantitative electron microscopy, in *Micromethods in Molecular Biology*, Vol. 14, Neuhoff, V., Ed., Springer, New York, 1973, 257.

466. Bahr, G. F., Carlsson, L., and Zeitler, E., Determination of dry weight in populations of submicroscopic particles by means of quantitative electron microscopy, *Int. Biophys. Cong. Stockholm*, 327, 1961.

467. Mazzone, H. M., Engler, W. F., and Bahr, G. F., Quantitative transmission electron microscopy for the determination of mass-molecular weight of viruses, *Methods in Virology*, Vol. 8, Maramorosch, K. and Koprowski, H., Eds., Academic Press, New York, 1984, 103.

468. Gaddum, J. H., Lognormal distributions, *Nature*, 156, 747, 1945.

469. Gaddum, J. H., Lognormal distributions, *Nature*, 156, 463,1945.

470. Bahr, G. F., Foster, W. D., Peters, D., and Zeitler, E., Variability of dry mass as a fundamental property demonstrated for the case of vaccinia virus, *Biophys. J.*, 29, 305, 1980.

471. van Dorsten, A. C., Oosterkamp, W. J., and lePoole, J. B., An experimental electron microscope for 400 kilovolts, *Philips Technical Review*, 9 (7), 193, 1947.

472. Cosslett, V. E., High voltage electron microscopy and its application in biology, *Philos. Trans. R. Soc. London B*, 261, 35, 197.

473. Dupouy, G., Electron microscopy at very high voltages, *Adv. Opt. Electron Microsc.*, 2, 167, 1968.

474. Dupouy, G., Performance and applications of the Toulouse 3 million volt electron microscope, *J. Microscopy*, 97, 3, 1973.

475. Dupouy, G., Megavolt electron microscopy, *Adv. Electronics Electron Phys. Suppl.*, 16, 103, 1985.

476. Hama, K., High voltage electron microscopy, in *Advanced Techniques in Biological Electron Microscopy*, Koehler, J. K., Ed., Springer, Berlin, 1973, 275.

477. Glauert, A. M., The high voltage electron microscope in biology, *J. Cell Biol.*, 63, 717, 1974.

478. Mazzone, H. M., Wray, G., and Engler, W. F., The high voltage electron microscope in virology, *Adv. Virus Res.*, 30, 43, 1985.

479. Maramorosch, K. and Sherman, K. E., Eds., *Viral Insecticides for Biological Control*, Academic Press, New York, 1985.

480. Martignoni, M. E., Breillatt, J. P., and Anderson, N. G., Mass purification of polyhedral inclusion bodies by isopycnic banding in zonal rotors, *J. Invertebrate Pathol.*, 11, 507, 1968.

481. Williams, R. C. and Backus, R. C., Macromolecular weights determined by direct particle counting. I. The weight of the bushy stunt virus particle, *J. Am. Chem. Soc.*, 71, 4052, 1949.

482. Polson, A., Stannard, L., and Tripconey, D., The use of haemocyanin to determine the molecular weight of *Nudaurelia cytherea capensis* B. virus by direct particle counting, *Virology*, 41, 680, 1970.

483. Wall, J. S., Hainfeld, J. F., and Simon, M. N., Biological scanning transmission electron microscopy, *EMSA Bull.*, 21, 2, 1991.

484. Wall, J. S., A High Resolution Scanning Microscope for the Study of Single Biological Molecules, Ph.D. thesis, University of Chicago, 1971.

485. Wall, J. S., Mass measurements with the electron microscope, in *Introduction to Analytical Electron Microscopy*, Hren, J. J., Goldstein, J., and Joy, D., Eds., Plenum Press, Corp., New York, 1979, 333.

486. Engel, A., Molecular weight determination by scanning transmission electron microscopy, *Ultramicroscopy*, 3, 273, 1978.

487. Smith, P. R., An integrated set of computer programs for processing electron micrographs of biological structures, *Ultramicroscopy*, 3, 153, 1978.

488. Trus, B. L. and Steven, A. C., Digital image processing of electronmicrographs: the PIC system. *Ultramicroscopy*, 6, 383, 1981.

489. Wall, J. S. and Hainfeld, J. F., Mass mapping with the scanning transmission electron microscope, *Annu. Rev. Biophys. Biophys. Chem.*, 15, 355, 1986.

490. Adolph, K. W. and Haselkorn, R., Comparison of the structures of blue-green algal viruses LPP-1M and LPP-2 and bacteriophage T7, *Virology*, 47,701,1972.

491. Mazzone, H. M., Engler, W. F., Wray, G., Gröner, A., and Zerillo, R., Electronmicroscopic analysis of a cabbage moth virus, *39th Annu. Proc. Electron Microscopy Soc. Am.*, 398, 1981.

492. Newcomb, W. W., Brown, J. C., Booy, F. P., and Steven, A. C., Nucleocapsid mass and capsomer protein stoichiometry in equine herpesvirus 1: scanning transmission electron microscopic study, *J. Virol.*, 63, 3777, 1989.

493. Mazzone, H. M., Engler, W. F., Wray, G., Szirmae, A., Conroy, J., Zerillo, R., and Bahr, G. F., High voltage electron microscopy of viral inclusion bodies, *38th Annu. Proc. Electron Microscopy Soc. Am.*, 486, 1980.

494. Bahr, G. F., Engler, W. F., and Mazzone, H. M., Determination of the mass of viruses by quantitative electron microscopy, *Q. Rev. Biophys.*, 9, 459, 1976.

495. Mazzone, H. M., Wray, G., Engler, W. F., and Bahr, G. F., High voltage electron microscopy of cells in culture and viruses, in *Invertebrate Systems in Vitro*, Kurstak, E., Maramorosch, K., and Dubendorfer, A., Eds., Elsevier/North-Holland, New York, 1980, 511.

496. Mazzone, H. M., Engler, W. F., and Bahr, G. F., Mass and molecular weight of bacteriophages T2 and T5, *Current Microbiol.*, 4, 147, 1980.

497. Williams, R. C., Backus, R. C., and Steere, R. L., Molecular weight determination by direct particle counting. II. The weight of the TMV particle, *J. Am. Chem. Soc.*, 73, 2062, 1952.

498. Oster, G., Doty, P. M., and Zimm, B. H., Light scattering studies of tobacco mosaic virus, *J. Am. Chem. Soc.*, 69, 1193, 1947.

499. Boedtker, H. and Simmons, N.S., The preparation and characterization of essentially uniform tobacco mosaic virus particles, *J. Am. Chem. Soc.*, 80, 2550, 1958.

500. Jennings, B. R. and Jerrard, H. G., Light scattering study of tobacco mosaic virus solutions when subjected to electric fields, *J. Chem. Phys.*, 44, 1291, 1966.

501. Doty, P. and Steiner, R. F., Light scattering and spectrophotometry of colloidal solutions, *J. Chem. Phys.*, 18, 1211, 1950.

502. Wall, J. S. and Hainfeld, J. F., Mass mapping with the scanning transmission electron microscope, *Annu. Rev. Biophys. Biophys. Chem.*, 15, 355, 1986.

503. Thomas, D., Newcomb, W. W., Brown, J. C., Wall, J. S., Hainfeld, J. F., Trus, B. L., and Steven, A., Mass and molecular composition of vesicular stomatitis virus: a scanning transmission electron microscopy analysis, *J. Virol.*, 54, 598, 1985.

504. De Blois, R. W., Uzgiris, E. E., Cluxton, D. H., and Mazzone, H. M., Comparative measurements of size and polydispersity of several insect viruses, *Anal. Biochem.*, 90, 273, 1978.

505. Welch, J. B. and Bloomfield, V. A., Concentration-dependent isomerization of bacteriophage T2L, *Biopolymers*, 17, 2001, 1978.

506. Chothia, C., Structural invariants in protein folding, *Nature*, 254, 304, 1975.

507. Smith, K. F., Harrison, R. A., and Perkins, S. J., Structural comparisons of the native and reaction centre cleaved forms of alpha 1-antitrypsin by Neutron and X-ray solutrion scattering, *Biochem. J.*, 267, 203, 1990.

508. Perkins, S. J., Smith, K. F., Amatayakul, S., Ashford, D., Rademacher, T. W., Dwek, R. A., Lachmann, P. J., and Harrison, R. A., The two domain structure of the native and reaction centre cleaved forms of C1 inhibitor of human complement by neutron scattering, *J. Mol. Biol.*, 214, 751, 1990.

509. Perkins, S. J., Nealis, A. S., Sutton, B. J., and Feinstein, A., The solution structure of human and mouse immunoglobulin IgM by synchrotron X-ray scattering and molecular modeling: a possible mechanism for complement activation, *J. Mol. Biol.*, 221, 1345, 1991.

510. Perkins, S. J., Chung, L. P., and Reid, K. B. M., Unusual ultrastructure of complement component c4b-binding protein of human complement by synchrotron X-ray scattering and hydrodynamic analysis, *Biochem. J.*, 233, 799. 1986.

511. Perkins, S. J., Hydrodynamic modeling of complement, in *Dynamic Properties of Biomolecular Assemblies*, Harding, S. E. and Rowe, A. J., Eds., Royal Society of Chemistry, London, U.K., 1989, 226.

512. De Blois, R. W. and Wesley, R. K. A., Sizes and concentration of several type C oncorna viruses and bacteriophage T2 by the resistive pulse technique, *J. Virol.*, 23, 227, 1977.

513. Tanford, C., *Physical Chemistry of Macromolecules*, John Wiley & Sons, New York, 1961.

514. Voelkl, E., Allard, L. F., Nolan, T. A., Hill, D., and Lehmann, M., Remote operation of electron microscopes, *Scanning*, 19, 286, 1997.

515. Chumbley, L. S., Meyer, M., Fredrickson, K., and Laabs, F., Development of a multi-user, networked scanning electron microscope, *Scanning,* (Abstr.), 18, 202, 1996.

516. Chand, G., Breton, B. C., Caldwell, N. H. M., and Holburn, D. M., World wide web-controlled scanning electron microscope, *Scanning,* 19, 292, 1997.

517. Hedges, J., Sarrafzadeh, S., Lear, J. D., and McRorie, D. K., Extension to commercial graphics packages for customization of analysis of analytical ultracentrifuge data, in *Modern Analytical Ultracentrifugation,* Schuster, T. M. and Laue, T. M., Eds., Birkhauser, Boston, 1994, 227.

518. Demeler, B., Ultrascan 2.5 Sedimentation Analysis Software, Exploration (Beckman Instruments) 1, 7, 1994.

519. Holzman, T. F. and Snyder, S. W., Applications of analytical ultracentrifugation in structure-based drug design, in *Modern Analytical Ultracentrifugation,* Schuster, T. M. and Laue, T. M., Eds., Birkhauser, Boston, 1994, 298.

520. Correia, J. J. and Yphantis, D. A., Equilibrium sedimentation in short solution columns, in *Analytical Ultracentrifugation in Biochemistry and Polymer Science,* Harding, S. E., Rowe, A. J., and Horton, J. C., Eds., Royal Society of Chemistry, Cambridge, U.K., 1992, p231.

521. Minton, A. P., Conservation of signal; a new algorithm for the elimination of the reference concentration as an independently variable parameter in the analysis of sedimentation equilibrium, in *Modern Analytical Ultracentrifugation,* Schuster, T. M. and Laue, T. M., Eds., Birkhauser, Boston, 1994, 81.

522. Hayes, D. B. and Laue, T. M., A graphical method for determining the ideality of a sedimenting boundary, in *Modern Analytical Ultracentrifugation,* Schuster, T. M. and Laue, T. M., Eds., Birkhauser, Boston, 1994, 245.

523. Shire, S. J., Analytical ultracentrifugation and its use in biotechnology, in *Modern Analytical Ultracentrifugation,* Schuster, T. M. and Laue, T. M., Birkhauser, Boston, 1994, 261.

524. Lewis, M. S., Shrager, R. I., and Kim, S.-K., Analysis of protein-nucleic acid and protein-protein interactions using multi-wavelength scans from the XL-A analytical ultracentrifuge, in *Modern Analytical Ultracentrifugation,* Schuster, T. M. and Laue, T. M., Eds., Birkhauser, Boston, 1994, 94.

525. Harding, S. E., Horton, J. C., and Morgan, P. J., MSTAR: a FORTRAN program for the model independent molecular weight analysis of macromolecules using low speed or high speed sedimentation equilibrium, in *Analytical Ultracentrifugation in Biochemistry and Polymer Science,* Harding, S. E., Rowe, A. J., and Horton, J. C., Eds., Royal Society of Chemistry, Cambridge, U.K., 1992, 275.

526. Philo, J. S., Measuring sedimentation, diffusion, and molecular weights of small molecules by direct fitting of sedimentation velocity concentration profiles, in *Modern Analytical Ultracentrifugation,* Schuster, T. M. and Laue, T. M., Eds., Birkhauser, Boston, 1994, 156.

527. Garman, E. F., Modern methods for rapid x-ray diffraction data collection from crystals of macromolecules, in: *Crystallographic Methods and Protocols,* Jones, C., Mulloy, B., and Sanderson, M. R., Eds., Humana Press, Totowa, NJ, 1996, chap. 4.

528. Brunt, A., Crabtree, K., Dallwitz, M., Gibbs, A., and Watson, L., *Viruses of Plants. Description and Lists from the VIDE Database,* Oxford University Press, New York, 1996.

529. Anderson, R., Mighell, A. D., Karen, V. L., Jenkins, R., and Carr, M. J., Electron diffraction databases, *Microscopy Soc. Am. Bull.,* 23, 128, 1993.

Index